CEPHALOPODS

Colin Dunlop and Nancy King

Cephalopods: Octopuses and Cuttlefishes for the Home Aquarium

Project Team
Editors: Craig Sernotti, David E. Boruchowitz
Indexer: Dianne L. Schneider
Design Layout: Stephanie Krautheim

T.F.H. Publications
President/CEO: Glen S. Axelrod
Executive Vice President: Mark E. Johnson
Publisher: Christopher T. Reggio
Production Manager: Kathy Bontz

T.F.H. Publications, Inc.
One TFH Plaza
Third and Union Avenues
Neptune City, NJ 07753

09 10 11 12 13 1 3 5 7 9 8 6 4 2
Printed and bound in China
Library of Congress Cataloging-in-Publication Data
Dunlop, Colin.
Cephalopods : octopuses and cuttlefishes for the home aquarium / Colin Dunlop and Nancy King.
p. cm.
Includes bibliographical references and index.
ISBN 978-0-7938-0658-4 (alk. paper)
1. Cephalopoda as pets. 2. Marine aquariums. I. King, Nancy. II. Title.
SF459.C39D86 2008
639'.7--dc22
2008022157

The Leader in Responsible Animal Care for Over 50 Years!®
www.tfh.com

CENTRAL Garden & Pet

Table of Contents

Dedication

To Ollie, who began it all, and to Erin, who is the future.

Acknowledgements

We would like to thank the following people who assisted us in various ways by providing help, support, and valuable information to make this book a reality:

Tony Morelli of TONMO.com.

We are grateful to those who participated in this book by reading sections, offering their photographs, and sharing their expertise and experiences: Greg Barord, Kat Bolstad, Roy Caldwell, Robyn Crook, Adam Daw, Phil Eyden, Christine Huffard, Jean McKinnon, Monty Montague, Richard Ross, Carol Sauer, Dale Tyler, and Denise Whatley. Special thanks to Roy, who shared his magnificent photos as well as his knowledge.

Also to our photographers, who helped make this book informative and visually interesting: Kevin Bylund, Dominick Cirigliano, Markus Holcomb, Dan Hyslop, Bill King, Matt McGhee, Dermot McMahon, Daniel Pon, Mike Roth, Mike Seidman, and Joe Strickland.

Friends and people who have helped over the years and would probably rather we never mentioned cephalopods to them ever again…

Andy Lister, Jim Wilson at M&R Dogfish, Cliff Murray at Worldwide Tropicals, Helen Gill of Simply Seahorses, Keith and Errin Todd of Aqualogistix, Andrew & Diane Stephenson of Educational Displays, Dermot McMahon, Charlie Howe, Matt Mitchell, Bob Reid, Ben Mitchell, and Allie Graham.

Special thanks to Adam Daw for information on the argonaut and his photos, Charles Delbeek for information on the bobtail squid, Kat Bolstad for her help and for providing photos of the giant squid, Richard Stride for supplying Colin with Sepia, and Janet and Gilbert Dunlop for just supplying Colin!

Thanks to Steve O'Shea and Roland Anderson for both information and inspiration.

To the people at the National Resource Center for Cephalopods who have shared their knowledge of breeding cephalopods with us during our visits over the years.

Our appreciation to Greg Ewald for creating an original Cthulhu painting for this book, and to Arthur Erickson for providing information about and a photograph of the octopus bag.

Nancy would like to thank Jay Hemdal of the Toledo Aquarium for his help and advice with the publishing process. Also, Michael Doan of Odyssey Pets for help with tank issues and Paul Sachs of the Aquaculture Store for information about fiddler crabs and cephalopod foods.

And the rest of the TONMO.com community, who helped us learn more about cephalopods over the years.

…and of course Mykela and Bill who have had to put up with us through this cephalopod obsession!!!

Preface

This book is the culmination of more than five years of planning.

Early in 2002 we first "met" on the forums of a cephalopod website called The Octopus News Magazine Online, abbreviated to TONMO.com. Both of us and site owner Tony Morelli all saw the need for a book on cephalopods, especially one covering how to keep octopuses and cuttlefishes in the home aquarium. At one point in 2002 we even got as far as writing the contents page and a few chapters before the project got shelved for further research.

Over the next few years we watched as this area of the aquarium hobby grew and the number of interested hobbyists increased exponentially. At the same time, we learned a lot more about keeping cephalopods and the ways in which we can keep our cephs longer and healthier. More recently, captive-bred cephalopods have started to become available thanks to the hard work of dedicated hobbyists, and there is now a greater chance that local fish stores are likely to be selling cephalopods. With this comes an even greater need for a comprehensive guide to keeping cephalopods.

This book is intended for the intermediate to advanced home aquarist and assumes at least some experience on your part in keeping a saltwater tank. It will give you the information you require to choose the right species and to provide a good environment for a healthy octopus or cuttlefish.

We think you'll enjoy the selection of photos, which were taken by not only us but also by many other cephalopod owners, fanatics, and scientists that we have had the great pleasure of getting to know over the years and from whom we have learned a lot. We have found their variety of imaginative approaches to cephalopod tanks interesting and inspiring, and we hope you will enjoy looking at the photos of the many octopuses, cuttlefishes, and other cephalopods that have shared our homes with us. To meet like-minded people and to see more of the setups we mention in the book, have a look at www.tonmo.com for yourself.

We know from contact with prospective cephalopod keepers that about 25 percent of first-time owners have had no prior saltwater experience, so we advise those people to buy a good introduction to saltwater aquariums, such as Michael S. Paletta's *New Marine Aquarium* (Microcosm Ltd./TFH Publications, Inc.). This book, or one like it, will ground you in marine aquarium husbandry, giving you a good foundation for keeping a cephalopod in the home aquarium.

We've also gone beyond cephalopod care to look back in time to see evidence of the importance of octopuses and other cephalopods in human history and to look at how important they are in today's culture.

We hope you'll enjoy reading this book as much as we enjoyed writing it, and we extend special thanks to all who contributed to it.

Colin and Nancy
Larkhall, Scotland and Dallas, Texas
April 2008

Chapter 1

Cephalopoda: History and Characteristics

Cephalopods (octopuses, squids, cuttlefishes, the vampire squid, and nautiluses) are mollusks, members of the phylum Mollusca. They are related to clams and oysters and snails and share certain features such as the radula, which is a tongue or feeding organ. Many mollusks can secrete a shell, and some of the cephalopods living today fall into this category.

CEPHALOPODS are found in ocean waters and in estuaries, but there are no freshwater species. This is interesting in itself, because many other marine invertebrates, such as worms, shrimps, lobsters, and snails, have freshwater equivalents. And of course there are no terrestrial cephalopods, despite the many jokes about and "photos" of the so-called tree octopus. There are, however, some cephalopods, such as intertidal octopuses, that can live for a number of minutes out of the water as they travel from tidepool to tidepool while searching for food.

The range of cephalopods in the oceans is much wider than most people believe. They are found in tropical waters but also in the Arctic and Antarctic oceans. Some are intertidal and are found near the shore in shallow waters, some live farther out, and still others are deep-water creatures. Some live near thermal vents. Certain cephalopods live in the deepest part of the ocean in permanent darkness. So cephalopods have developed in a way that they occupy many niches and are found in some form in almost all ocean waters, but not all of them necessarily in any one place. For example, the United States has no cuttlefishes off its shores and no octopuses along the shores of the New England states.

Octopuses, Octopi, or Octopodes: Which Is the Correct Plural?

Sometimes you will see octopuses referred to as "octopi," or even "octopodes," instead of the preferred plural "octopuses." Many people assume that the word "octopus" comes from the Latin and therefore should follow Latin rules for plurals: "octopi." This is wrong. The word "octopus" actually is a Latinization of the Greek *oktopous* (οκτωπους), signifying "eight" + "foot." The plural of that would be *octopodes* (οκτωποδες). The commonly accepted plural is simply "octopuses," which is what we use in this book. Interestingly, if the word were Latin the singular would be *octopes* ("eight" + "foot"), with the plural being *octopedes*.

THE LIVING CEPHALOPOD GROUPS

Before we continue further, you should know that when we discuss living cephalopods in this book we are referring to octopuses (including the argonaut), cuttlefishes, squids, and nautiluses. However, because most home aquarists keep only octopuses and cuttlefishes, we will concentrate primarily on those two groups, with a small section for nautiluses, squids, and the argonaut. (See Chapter 8.) We're not dealing at all with the other living cephalopod group, the order Vampyromorpha, which contains only one species, *Vampyroteuthis infernalis*, the vampire squid.

ANATOMY, CAMOUFLAGE, AND REPRODUCTION

The cephalopods, with the exception of nautiluses, share similarities in their life cycles, their growth, and their reproduction. They are short-lived, grow quickly to reach sexual maturity, and reproduce at the end of their lives. Most have just a single breeding cycle and die shortly thereafter.

Cephalopods are the most intelligent of all invertebrates and have relatively large brains—larger than other mollusks—as well as highly complex nervous systems.

How Many Cephalopods Are There?

More than 700 cephalopod species living today have been identified, and there are many more yet to be described.

Cephalopods Are Predators

Cephalopods are all fast-growing, relatively short-lived predators. They have a hard chitinous beak, much like a parrot's beak, and a sharp tongue called the radula. Octopuses can use this tongue for drilling holes into the shells of bivalves.

Anatomy

The term "cephalopod" is formed from two Greek words meaning "head" and "foot" and refers to how cephalopods are anatomically constructed, with the arms, and in some cases tentacles, coming directly off the head surrounding the mouth. In the case of octopuses, cuttlefishes, and squids, the body organs are contained in the mantle, a bag-like structure that most people just assume is part of the head. The head (where the eyes are situated) of the cephalopod is actually between the arms and the mantle.

Look carefully. Can you find the octopus? Here's a hint: Look along the upper right edge of the rock.

Basic Physiology

These are the most significant external features of octopuses and cuttlefishes.

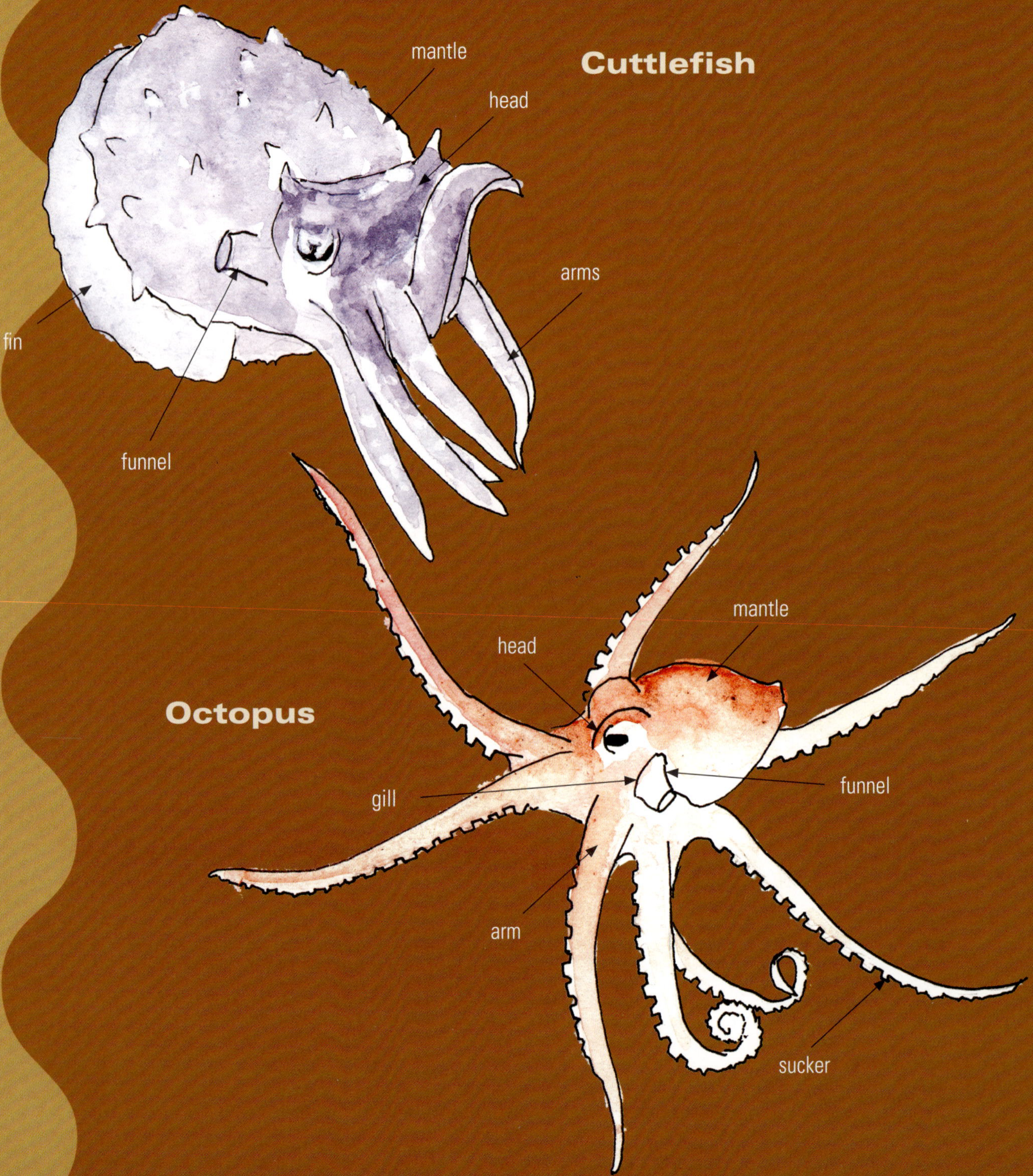

Uses of Cuttlebones

The cuttlebone is the internal shell of cuttlefishes and is a by-product when the cuttlefish is caught for food. It is familiar to keepers of birds, who offer cuttlebones to their pets as a source of trace minerals and calcium, and to help keep the bird's beak sharp. Cuttlebones are also sometimes used as a mold for metal casting.

Cephalopods have gills, two to four openings used to pump in sea water, which they expel through the funnel (sometimes called the siphon). Nautiluses have four gills; the other more advanced cephalopods have only two. There is only one funnel, which can make an octopus appear asymmetrical.

Cephalopods are valued in medical research because of the large nerve fibers in the mantle, which makes them easier for neurophysiologists to study. Usually squids and cuttlefishes are chosen for this kind of study, while octopuses are more often studied for behavior. However, octopus arms have recently been admired and studied by robotics experts, and octopus eyes have recently been studied as well.

Cephalopods have acute vision and sophisticated eyes. Octopuses, cuttlefishes, and squids have a lens, while nautiluses have an eye more like a pinhole camera. Octopuses have excellent vision, and they can distinguish size, shape, and even polarization of light. It is thought that they are most likely color blind, which will amaze anyone who has seen them match their surroundings exactly.

Cephalopods have copper-based blue blood (in contrast to our iron-based red blood). Surprisingly, cephalopods, like most other invertebrates, are very sensitive to copper in the sea water, and it can cause their death. This is why home aquarists have to be very careful about buying used tanks; if copper-based medications were used in a second-hand tank the residue can be lethal.

There One Minute and Gone the Next

Most cephalopods, especially octopuses and cuttlefishes, are excellent at camouflage. Squids can also change colors, and even argonauts are capable of some color change. The ability can vary by species and affects both color and texture. These cephalopods use chromatophores, certain color cells, to change pattern and brightness and have additional types of cells (iridophores and leucophores) to reflect light from the environment.

They can raise patches of skin to produce texture and can do this in such an elaborate way that some octopus species can resemble seaweed. Octopuses and cuttlefishes can blend in or contrast with their surrounds and can display mating patterns in an instant. Cuttlefishes can match a pattern when placed over a patterned surface. And some squids can use a light organ to make them appear invisible to predators. Some squids can even display half-and-half color across their body, especially at mating times—one half to warn off other males and one half to woo females—swapping sides in the blink of an eye if need be.

The ability to change color, brightness, patterns, and texture is part of the defensive arsenal of octopuses, cuttlefishes, and squids. In addition to making themselves hard to find through immediate changes in appearance, those cephalopods can use an ink-like substance to discourage and stop predators and give themselves a chance to escape.

Most have an ink that they can squirt through the funnel as a defense. This dark-colored (actually concentrated red) ink is composed of melanin and sometimes mucus; it can be diluted with water. Most octopus and cuttlefish keepers will experience inking, which can come in two ways: as a pseudomorph—thicker ink to make a shape similar to that of the cephalopod in order to divert a predator while the cephalopod escapes—or a general smokescreen, which is what we most often see in our tanks. The ink contains an irritant, tyrosinase, which can further hinder any predator by irritating its eyes and maybe even affecting its sense of smell.

Cephalopods can also avoid predators by jetting away (or retreating into a shell, in the case of nautiluses). They can use their funnel to squirt water, but this is usually to fend off an annoyance rather than a predator.

Reproduction

Reproduction by various cephalopod species will be discussed throughout the book. In general, the male has a packet called the spermatophore where he stores his sperm. There may be displays of courtship or interaction, followed by copulation. The male usually places the spermatophore into the mantle cavity of the female by using the hectocotylus, which is an arm specialized for reproduction. The female can delay laying eggs until she is ready. She usually attaches the fertilized eggs in clusters to a hard surface, such as the rocks that make up her den. The reproductive phase usually comes at the end of an octopus's life.

Typically males and females do not live very long after reproducing. This means that keeping a cephalopod is not for the faint of heart. Many keepers develop strong bonds with their octopuses or cuttlefishes, and the short cephalopod lifespan is difficult for some home aquarists to cope with. Others go on to keep octopus after octopus, or generations of cuttlefishes, and are successful at breeding their cephalopod of choice.

WHY CEPHALOPODS INTEREST US

Cephalopods are highly evolved organisms and are regarded as the most intelligent of the invertebrates. This makes them interesting for home aquarists because of the possibility of observing intelligent behavior, especially in the case of the octopus, as well as having chances to interact with them.

Can I Write With That Ink?

Is cephalopod ink like pen ink? The answer is no; these two inks are completely different. Our present fountain pen ink is made of synthetic dyes and compounds, while cephalopod ink is composed mostly of melanin. However, cephalopod ink, especially cuttlefish ink, has been used historically in the brownish artists' ink known as sepia. We actually get the name of the color sepia from the ink of the cuttlefish *Sepia officinalis*. The brown sepia ink was popular in the nineteenth century for drawings and was made by mixing shellac with dried cuttlefish ink.

Ollie and Nancy Hunt Crabs Together

One author of this book, Nancy, kept an *Octopus bimaculoides* and liked to find interesting ways to interact with her. Nancy discovered that when she dropped a couple of fiddler crabs into the water, they hid in the sand. Ollie didn't always see where they landed and didn't know where to look. So Nancy started pointing them out to Ollie, using her index finger on the outside of the aquarium glass to point at the hiding treats on the inside. Ollie soon picked up on the meaning of the finger and hurried over to dig for the crab in that spot. In this way Ollie and Nancy hunted crabs together.

In the last few years many home aquarists keeping octopuses have experimented with having their octopus open jars or other containers to find a treat inside. There are many ways to play with your octopus, such as tug of war with the feeding stick, or even a sort of hide and seek. Your octopus can be observing you at the same time you're observing it. Octopuses make eye contact, unlike most other invertebrates.

CEPHALOPODS OF THE PAST: THE RECORD IN FOSSILS

If you look at cephalopod fossils, you'll find a complicated history and development. It is not a straight line from the oldest fossils to today's modern octopuses and cuttlefishes. The exception is the nautilus, which is the oldest cephalopod subclass. The nautilus is often referred to as a living fossil, because it apparently does not differ greatly from nautiloids in the Cretaceous Period (144 to 65 million years ago). Even contemporary nautiluses are distinguished from the rest of the cephalopods by their long lifespan and reproduction over the course of their lives.

The fossils fall into three taxonomic subclasses:

- Nautiloidea, a straight or spiral external shell
- Ammonoidea, usually a spiral external shell
- Coleoidea, a straight internal shell

The first tiny cephalopods appeared about 510 million years ago (Late Cambrian), having developed from more primitive mollusks. These early cephalopods, subclass Nautiloidea, developed a shell with a series of chambers, somewhat like what the nautilus has today, and found a way to control buoyancy. This chambered external shell was great for defense, as the cephalopods could retreat into the shell for protection. Most of these primitive nautiloids had long cone-shaped shells, unlike the spiral shape of todays' nautiluses. Some of these animals grew to be huge in size—one such nautiloid called *Cameroceras* grew to approximately 36 feet (11 m) in length and was the fiercest creature in the oceans of the time.

Ammonites were named after Amun, an Egyptian deity who was sometimes shown having the head of a ram with curved horns. This corresponds to the coiled shape of an ammonite. The first primitive ammonoids appeared in the Early Devonian, about 390 million years ago, having evolved as an offshoot of the nautiloids. True ammonites are first recorded from the Early Jurassic and persisted right through the Mesozoic. They rapidly became the dominant cephalopods largely replacing the once abundant nautiloids, which in turn became greatly reduced in variety and number. Over a period of

This ammonite, *Cleoniceras besairei*, has been half polished to reveal the lines of the sutures where the individual internal chambers met the shell wall. They leave beautiful leaf-like patterns that are unique to each ammonite genus and help experts identify them. The unpolished half shows the original shell. When alive this would have probably been brightly colored and may have shown complex designs on the shell. The ammonite hails from Tulear, Madagascar, from about 105 million years ago.

almost 330 million years, ammonoids evolved rapidly, diversified and became increasingly complex, but ultimately were not successful, becoming extinct slightly after the dinosaurs, around 65 million years ago. Ammonites came in many shapes and sizes, some huge, growing over 6 feet (1.8 m) in diameter, while others developed vertically spiraled, straight, spiked, or even hooked shells.

Subclass Coleoidea comprises cephalopods that have lost their external shell but usually have remnants of this shell internally. They are the successors of the ammonites. It was in the Jurassic Period (208 to 146 million years ago) that the belemnites (an extinct group of cephalopods) as well as the first squid evolved. Given that hard structures are much more likely to fossilize, and because many cephalopods are totally soft bodied, it's not surprising that only a few octopus and other coleoid fossils survive today. We don't know in great detail how they evolved. However, octopuses are found as far back as the Carboniferous (360 to 300 million years ago). The oldest known specimen is a single fossil of a ten-armed

An octopus displaying deimatic behavior, making itself look larger and more menacing.

octopus called *Pohlsepia* that was recovered from the famous Maxon Creek ironstones in Illinois and is 296 million years old. We know even less about the cuttlefish, but it is thought that they probably first appeared early in the Cenozoic after the dinosaurs became extinct. However, fossils of these animals are so rare that even paleontologists cannot agree on the dates.

Not having an external shell meant that new forms of cephalopods had to rely on other means of defense: camouflage or jetting away to avoid predators. They did not have the buoyancy or the protective shell of their ancestors to rely on—rather, they had to depend on intelligence and mobility.

Existence and Distinction

From the many cephalopod (class Cephalopoda) forms that developed in geological time, only two of the three major groups are extant today: subclasses Nautiloidea and Coleoidea. The former comprises a few nautilus species, and all the remainder of today's species fall into the latter.

The nautiloids, as we've mentioned, have an external shell with chambers, and the other modern cephalopods have reduced or lost the shell almost completely. Squids have a reduced shell without chambers, so they have no buoyancy, and octopuses have developed even further and have no shell. Another significant difference is found in the heart and gills. While the nautiluses have one heart and four gills, the cephalopods belonging to Coleoidea have two gills and three hearts (one central and one at each gill).

Cephalopod Fossils

Cephalopods have existed for many millions of years. At one time they were top predators in many habitats. Although there are still many species, there used to be many more. Some of these are discovered and studied when their fossils are unearthed. Below you will see just a small sampling of cephalopod fossils. Each is numbered, and a description of the fossil follows on the next page.

(1) Ammonite, no details known.

(2) Belemnite, *Belemnitella mucrata*, 70 million years ago (Maastrichtian, Late Cretaceous), Sussex, UK. Internal guard (pen) preserved in chalk.

(3) Ammonite, *Otoites sauzei*, 170 million years ago (Bajocian, Jurassic), Neuquen Basin, Argentina. Unusual globular ammonite.

(4) Ammonite fragment, *Baculites* sp., 73 million years ago (Lower Maastrichtian, Late Cretaceous), Pierre Shale, Colorado. Section of straight shell of a very late uncoiled ammonite. This would have lived head down with the shell pointing straight up.

(5) Ceratite ammonoid (Triassic), Timor. Species and precise date not known.

(6) Nautiloid, *Orthoceras* sp., 440 million years ago (Ordovician), China. Polished specimen. Very common in museum souvenir shops.

(7) Nautilus, no details known (probably Cretaceous, Madagascar).

(8) Ammonite, *Desmoceras latidorsatum*, 100 million years ago (Albian, Early Cretaceous), Madagascar.

(9) Ammonite, *Cleoniceras besairei*, 105 million years ago (Albian, Early Cretaceous), Madagascar.

(10) Ammonite, *Perisphinctes* sp., 156 million years ago (Oxfordian, Jurassic), Madagascar.

(11) Nautilus, 100 million years ago (Albian, Early Cretaceous), Madagascar. Species not known.

(12) Ammonite, *Dimorphoplites* sp., 100 million years ago (Albian, Early Cretaceous), Folkestone, Kent, UK.

(13) Ammonite, no details known.

(14) Nautiloid, *Lituites* sp., 460 million years ago (Ordovician), Hupei, China. Fragmentary specimen.

(15) Ammonite, *Perisphinctes* sp. (Oxfordian, Jurassic), Madagascar. Displays lappets on aperture indicating mature specimen.

(16) Ammonite, *Desmoceras latidorsatum*, 100 million years ago (Albian, Early Cretaceous), Madagascar.

(17) Goniatite ammonoid (Devonian). Species and location not known. This early ancestral ammonoid has simple suture lines demonstrating its primitive nature.

Chapter 2

Planning to Keep an Octopus or Cuttlefish

The responsibility of looking after any type of living thing is not something to take lightly. A captive animal relies on you for everything it needs to stay alive, to stay healthy, and to have a life that is both enriched and safe. You are going to be in charge of that animal's aquarium and its life-support system, and if you make a mistake it can cost your cephalopod its life.

NOT FOR BEGINNERS

Cephalopods are not for beginners. They require a stable aquatic environment free of pollutants and an aquarium set up specifically in a manner that will suit them. It can be quite costly at first to get all the necessary equipment, but that cost is justified in terms of the lives of the cephalopods and other animals it will save.

Know Before You Go

Hobbyists who are totally new to the marine aquarium hobby should avoid an octopus or cuttlefish, at least for a few months to a year to get all of the beginner mistakes and learning curves out of the way. You should set up a tank similar to a reef environment and keep hardy fish and invertebrates for a while and get used to reading water quality, doing water changes, etc. We aim the challenge of cephalopod husbandry at intermediate to advanced aquarists.

Aquarists who are familiar with the nuances of caring for marine invertebrates will still discover challenges, but their experience with and understanding of equipment and water parameters will go a long way towards maintaining a thriving setup. Tank equipment and water quality should not be an enigma by the time you decide to keep cephalopods. You need to be able to maintain a high-quality and stable marine aquarium environment.

Octopuses and cuttlefishes are not good beginner animals. They are best left to intermediate or expert marine hobbyists.

Octopuses can survive out of water for a short period of time before they dry out, like this *Octopus vulgaris* foraging in a shallow coastal rockpool in South Africa.

They Need Live Foods

Caring for a cephalopod is going to require the use of live food. No cephalopod species have been reared exclusively on dead food. They may be coaxed into accepting substitutes for live foods somewhere along the line, but there is no escaping the fact that at some point in their lives—maybe even the major portion of their lives—you are going to have to provide live food. Some people may have an issue with feeding living animals, regardless of species, to another animal, but there is no way around this. Prospective keepers have to make their own decision. It will certainly give some people something to think long and hard about, and nothing in this book will try to convince you either way. Unfortunately, live foods are a necessity for keeping cephalopods, especially in their early weeks and months.

Gone Too Soon

One last point that will certainly hold some meaning with experienced cephalopod aquarists is that possibly the worst thing about keeping any octopus or cuttlefish is that they have such a short lifespan. It's sad but true that most cephalopods seen in the hobby have a natural lifespan of only a few months, some as short as six months and the longest not much over a year. It's unusual for an invertebrate to become as endearing a pet as many octopuses can, and it always seems such a shame that they don't live longer, especially when so many become a part of the family. For this reason some home aquarists keep an octopus in their tank at all times, replacing one that dies with a younger octopus. Another solution

How to Spend Your Time

If you want to keep cephalopods you should spend a lot of time reading about the set-up and maintenance of marine aquaria. This can be done by buying books like the ones mentioned in this book's bibliography, or by joining an online forum for saltwater fishkeepers or reefkeepers. Most of what is needed for reef aquaria relates to cephalopod aquaria, so in this regard there is a lot of information out there. A good website to join is The Octopus News Magazine Online at www.tonmo.com. Here you will get the most up-to-date cephalopod information and advice.

You will also need to commit a lot of time to maintaining your aquarium. In later chapters we will provide lists of daily, weekly, and monthly tasks that you will need to perform.

Lastly, you may need to reorganize that summer vacation! If you are going to keep an octopus, you may have to forfeit leaving home for any longer than two days at a time. Unlike as with most fishes, there can be no lapses in feeding for octopuses! Cephalopods need attention on a daily basis.

that makes the short lifespan more tolerable is to keep more than one octopus tank at a time. The silver lining here is that by providing an ideal environment for your cephalopod you can know that it was well cared for through its entire life.

COMMITMENTS OF TIME, MONEY, LEARNING

A new aquarist is about to have a very steep learning curve if starting from scratch down the road to keeping a cephalopod. There are commitments to be made—the three biggest are money, learning, and above all else time. This is not an instant gratification hobby; wanting to keep an octopus on a whim because it's "cool" will normally end in tears. As moderators on a busy cephalopod website, we see people turning up for advice constantly after an impulse purchase. This has even included one new member returning from an aquarium store with a tank, equipment, salt water, *and* an octopus in a bag, all in one day! Again we emphasize that prior aquarium experience and a bit of patience are a necessity. You *must* be patient. There are no quick fixes with keeping cephalopods, and setting up their aquarium includes a cycling period of at least three months.

It's important to thoroughly understand the requirements of the octopus or cuttlefish you're considering buying. Read and learn as much about the species as you can. It's helpful to look at cephalopod websites and read about actual ceph-keeping experiences. Also read about where the particular species you are interested in lives and in what sort of environment (reef, along the shore, in rocky areas, for example).

CAVEATS

Keeping cephalopods isn't always going to be a smooth ride. These are not particularly easy aquarium animals. Right from hatching they require a specialized diet, exact water quality, expensive equipment, and lots of your time. And there's that short lifespan. With cephalopods it certainly is a matter of live fast, die fast.

REWARDS

If you are up for the challenge, you are going to be rewarded with one of the best experiences a home aquarist could ever hope for. You are going to have the world's most intelligent invertebrate sharing your living space, one that can change color faster than you can blink, one that can change its body's shape, one that can learn tricks like opening jars—and above all just be a cephalopod doing all the cool stuff that cephalopods can do!

SETTING UP A CEPHALOPOD TANK

As you must have already gathered, keeping a cephalopod is not as simple as filling a tank with salt water and throwing in an octopus! The cephalopod aquarium has to be set up several months before the

The common octopus, *Octopus vulgaris*.

animal's arrival to allow the aquarium's biofilter to mature. The biofilter's maturation process is not an exact science and is something that has to be monitored very closely. Introducing a cephalopod too soon into an aquarium that is not mature can be lethal.

You should use the biofilter's maturation time as an opportunity to create the ideal habitat for the animal you want to keep. You should research its natural habitats and its life cycle; most of all, if you have never kept a marine aquarium before, get used to doing regular water changes, testing water conditions, and everything else you must to do to maintain a marine aquarium. Mistakes will be made, so make them early on—before you have an octopus in the tank.

Choosing a Cephalopod For Your Home Aquarium

Among the known cephalopods, only octopuses and cuttlefishes are commonly kept in home aquariums. Other cephalopods, such as squids and nautiluses, are found in research labs and public aquariums.A few individuals have kept nautiluses, and once in a while someone tries to keep a small squid they've found along the shore. A very few people find argonauts in the ocean or washed up on the shore and try to keep them. We will provide information on keeping all of these cephalopods, but we'll emphasize octopuses and cuttlefishes, which are available to the pet trade and sometimes bred by hobbyists. Of these two, octopuses are far more commonly kept, at least in the United States.

What to Take Into Account

Some of the most important aspects to consider are the availability, the eventual adult size, the age, the temperament, and even the conservation status of each cephalopod species.

Availability

Octopuses are more widely available in the pet trade than cuttlefishes. But be aware that various species come onto the market from time to time and that not all are correctly identified. "Common octopus" should refer to *Octopus vulgaris*, but in fact normally means "any old octopus." Likewise, many dealers will label any old octopus with the name "*Octopus vulgaris*" without really knowing the proper identification. There are hundreds of octopus species, so it's possible you could get one of the more uncommon ones. It is possible to have good experiences with any octopus or cuttlefish as long as you adapt to the cephalopod you bought, but do your best to properly identify the species that interests you.

Cuttlefishes can change colors almost instantly to hide from both predators and prey. This one has taken the color of its tank's substrate.

Age

As we said earlier, cephalopods have a short lifespan. Therefore you should do your best to find out how old the specimen you want buy is. The cuttlefish *Sepia bandensis* is sometimes available in the form of eggs, giving you the chance to raise cuttles from larvae. Octopuses are sold at anywhere from one month old to fairly mature. Often octopuses are wild caught, but sometimes tank-bred octopuses are offered for sale. You can be surer of an octopus's age and species by buying a tank-bred specimen.

Adult Size

The eventual adult size that a cephalopod will reach is also an important consideration. The cuttlefish *S. officinalis* occasionally comes onto the aquarium market but needs a very large tank. If you can provide the right environment, this species is lively and interesting to keep. However, most home aquarists choose *S. bandensis*, which in captivity may grow to a total length of only 6 inches (15 cm).

As for octopuses, *O. vulgaris* may be one of the most interesting species to keep, but it too requires a very large tank. Many sizes of octopus are available, from dwarfs that will live in a 30-gallon (120-liter) tank to mid-sized octopuses such as *O. bimaculoides,* better suited to 50 gallons (200 liters) and upwards.

Most available species will live at tropical water temperatures (78°F, 26°C) but a few, such as the California two-spot octopus, *O. bimaculoides*, and the cuttlefish *S. officinalis,* need cooler water and could require a chiller. One of the smaller cuttlefishes from warmer waters, such as *S. bandensis*, and the octopus *Abdopus aculeatus,* make fewer demands on tank size, and it's not difficult to keep the water at 78°F. Both species would be comfortable in their own 50-gallon tanks.

One or Many

There are differences in requirements for keeping octopuses and cuttlefishes. For instance, octopuses may try to escape, so the tank must be made escape-

Go Small

Don't have room for a big cephalopod setup? Dwarf octopuses enable you to have the experience of keeping an octopus without having to buy a large tank, but they present their own problems. *Octopus mercatoris* is readily available but is nocturnal, meaning that you'll mostly see it only when the lights are off in the tank. If you don't mind using a red light, these little octopuses can be interesting to watch, since they do not see red light and will come out and behave normally while illuminated this way.

Unlike most other species, if raised together *O. mercatoris* individuals can successfully be kept in multiples, even in a 30-gallon (120-liter) tank with minimal escape-proofing. This is a large-egg species that can be captive raised. Home aquarists who keep these little octopuses enjoy having them but admit that, as a whole, they're not as interactive as larger diurnal species and, if wild caught, are likely to be full-grown adults with little natural lifespan remaining.

Captive-Bred Versus Wild-Caught

We as aquarists have a responsibility to the animals that interest us. We should be helping in the conservation of cephalopods and in advancing the available information on them. One of the areas in which we can make the cephalopod hobby more sustainable is by keeping captive-bred (CB) animals.

Captive breeding can take the strain off wild stocks by ensuring that wild octopuses are left in the sea. No species of cephalopod currently is listed as being endangered, but this could possibly change soon as more is learned about their natural history.

Pros and Cons of Owning an Octopus

Owning an octopus can be a remarkable and enjoyable experience, but it has its downsides, too:

Pros

- They are intelligent and will interact with you and their environment.
- They display interesting color and skin texture changes.
- They can solve problems, such as opening jars.
- Many species are available to suit different tanks.
- They can usually be convinced to take dead food after a while.

Cons

- Food can be expensive, and a steady supply of live food can be difficult to obtain.
- They will escape if you do not properly seal the tank.
- They have a short life span—painfully short for an interactive pet.
- They are optimally kept alone rather than with other animals.
- Often it is difficult to get the species you really want.
- Some species ship poorly.
- Some species are toxic to humans.
- Some exotics should be avoided for conservation reasons.

proof. Cuttlefishes are often kept in groups, while octopuses are usually alone in a species tank. Can you provide a good home for a group of cephalopods, or are you content with just one? Decide ahead of time.

CEPHALOPOD TANKMATES

This is not an easy subject. We are of the opinion that cephalopod tanks should be set up as a species-only tank. This is especially true with octopuses. Although it can be tricky with cuttlefishes, it is possible to keep those cephalopods with a few motile invertebrate species and those that have hard protective shells like hermit crabs and most snails. The two main reasons for making a one-animal-only system are: (1) an octopus will try to eat almost any other animal put in with it, and (2) the cephalopod aquarium needs a lot of filtration and water quality management—adding more livestock increases the biomass and therefore the load on the filter.

Octopus mercatoris in a typical defensive position.

Other Cephalopods

Some cuttlefishes can be kept in groups, but there really shouldn't be any need to keep other animals in with an octopus. Octopuses are more like a personable pet than a tank full of fishes, and they present enough of their own peculiarities to keep you occupied. Any prospective octopus keeper is advised to keep only one octopus per aquarium, unless keeping a species like *O. mercatoris*, which is known to live successfully together in a small group of similar-sized animals. There really aren't any tankmates that are tried and tested as being proven safe to keep with octopuses. More advanced keepers or public aquariums with very large tanks may be able to permanently house other species of animal with their octopuses, but

Captive Bred Is Better

You will know the exact age of a captive-bred (CB) cephalopod, which is especially important when keeping extra-short-lived species. The animal will be used to aquarium conditions and will settle into tank life much more quickly than a wild-caught animal, and it will be accustomed to an aquarium diet, which will facilitate a quicker settling. The CB cephalopod will not have been subjected to the same capture and traveling stresses as a wild-caught animal, and no wild stocks will have been decreased. Also, CB animals are almost certainly free of diseases and parasites. Unfortunately, very few people are breeding cephalopods, so finding a CB specimen will be difficult. Additionally, not every species can be successfully captive bred at this time.

it is often difficult and requires a certain amount of luck, especially if having more than one member of the same species. There are well-documented cases with video footage of giant Pacific octopuses catching and eating sharks within a public aquarium! Always bear in mind that large octopuses eat fishes and large fishes eat octopuses. The predator-prey relationship often depends only on relative size.

Octopuses are by nature inquisitive animals and will explore every item within their tank in great detail. In its most basic form the octopus is trying to decide whether the object is something it can eat or something it can use. Rocks and pipes will be moved around to make dens, and every item of tank equipment is at risk of this thorough examination. So, too, is any other living thing within the aquarium.

There are two criteria that any species of animal you might want to add to a cephalopod's aquarium must fulfill: it can't harm the cephalopod and the cephalopod can't harm it. Being predatory in nature, some cephalopods live on a diet that includes other mollusks (including other cephalopods), crustaceans, and fishes. Asking multiple cephalopods to cohabit in the same tank is similar to mixing lions and gazelles in the same zoo enclosure.

Some cuttlefishes will sometimes cohabit with conspecifics, but even this can lead to fights and cannibalism. As far as octopuses are concerned, it would seem that *O. bimaculoides* is more of a gregarious

Sarcophyton leather coral is safe with a cephalopod.

species and will tolerate other octopuses in very large aquariums. By comparison, *O. vulgaris* will constantly harass any other cephalopod. The danger in mixing tankmates is very much species-specific and can even vary from one individual to another.

Sessile Invertebrates

This is the group of ocean dwellers most likely to offer suitable tankmates, specifically corals and zoanthids. Only the most robust corals, such as the *Sarcophyton* leather corals, are likely to be able to withstand the conditions within a cephalopod tank. It is very important not to include the coral species that could easily sting and injure an octopus—species such as *Millepora* should be avoided at all costs.

Millepora fire coral. Do *not* put this coral into your cephalopod tank.

Corals and other sessile invertebrates are at risk in an octopus tank at least partly because of the octopus's redecorating skills. It is not out of the question that you may find your corals or zoanthids buried under a pile of rubble.

Sessile invertebrates will fare much better in a cuttlefish aquarium. In fact you can set the aquarium up as a reef tank and the cuttlefish will appreciate the rocky hiding places. Remember to avoid stinging species. It is also good practice to add the cephalopod last to make sure the other animals are established.

Motile Animals

Some people have had luck mixing fishes and octopuses, but it can be tricky. For example, damselfishes that have been used to cycle a new aquarium are frequently left in the tank (often because it proves impossible to remove them without disassembling the tank). Sometimes the octopus will get lucky and catch and eat the damsel. However, in other cases the damsel will attack the octopus and bite it at every chance, resulting in an injured and scared octopus that always hides. Things can start out smoothly, but sooner or later they'll probably turn nasty.

Trying to keep cuttlefishes with true fishes is definitely not a good idea. Cuttlefishes hunt in a manner

Even venomous lionfishes aren't safe from cephalopod predation.

that allows them to catch and eat fishes much more easily than octopuses can. Fish should be left out of cuttlefish tanks altogether. Even venomous species are not safe—cuttlefishes will eat lionfishes of the genus *Pterois* for example.

There are a lot of motile invertebrates, especially those with hard shells such as hermit crabs, urchins, and most snails, that can co-exist with a cuttlefish, but these same animals will be eaten by an octopus. Table 1 on the next page will provide you with an idea of appropriate and inappropriate tankmates for your cephalopods.

Species That Can Be Kept Together

Octopus mercatoris has been successfully kept in a group of six or seven and will reproduce under these circumstances in the home aquarium. Reasonable success has been seen with keeping other individuals of the same species such as *O. bimaculoides* and *Eledone cirrhosa* in large tanks. These species have had successful matings with resulting offspring. Many other species are territorial and cannot be kept in the same tank, except under observation for mating.

Tankmate Compatibility (Table 1)

Species	Octopus-Safe	Cuttlefish-Safe
Fish	Depending on species and with extreme caution	NO
Hermit crabs	NO Will be eaten by younger/smaller octopuses	YES
Brittle starfish	YES	YES
Snails	NO Will be eaten	YES
Corals	YES if avoiding sensitive species or those with potent stings	YES if avoiding sensitive species or those with potent stings
Anemones	As with corals	As with corals
Sea cucumbers	NO	YES
Shrimps (including cleaners)	NO Will be eaten	NO Will be eaten
Crabs	NO	NO
Bristleworms	YES	YES
Feather duster worms	NO (can get buried under rubble and pulled about)	YES with care
Other cephalopods	NO	With care and assuming a similar size and species

Note: The tank cleaning crew (crabs and snails, at least) is often eaten by an octopus when the octopus is young. As the octopus grows, however, it will have less interest in small invertebrates, so it may be possible to replace them at that time.

Multiple Octopuses?

Under certain circumstances two octopuses can be kept together. Unlike cuttlefishes, who will ignore each other for the most part and can be kept with others of the same species, octopuses are more of a challenge. Before you try to keep two octopuses in the one tank at the same time you should first ask why you want to do this and whether your tank can really accommodate two or more. If you are just starting out in cephalopod husbandry, we recommend that you keep octopuses singly, as you are going to have your hands full for the time being.

Not enough is known yet about keeping multiple octopuses in the same aquarium to really give definitive do's or don'ts, but here are some points to consider:

- Is your tank and system big enough to house them?
- Many octopuses will prey on other octopuses in the wild. Do you have a species that will not do this?
- Most octopuses set up a territory and defend it. Do you have enough suitable sites for each octopus to have its own space?
- The octopuses may be tolerant of their own kind as juveniles but much less so as they become adults. Are you prepared for this?
- Do you have another aquarium properly set up if you need to split your octopuses up in a hurry?

The next stage for keeping octopuses in captivity is to be able to produce captive-bred stock rather than relying on wild-caught animals. The only way that future generations can be born is by keeping two octopuses together for mating purposes. Unfortunately, captive breeding is still largely an unknown, and as equally unknown are successful documented reports of multiple specimens cohabitating.

So if you have successfully kept a few individual specimens and think your system can handle a pair, then carefully and with full consideration give it a go. Just make sure to report your findings!

SPECIES TO AVOID

You should not try to keep any species of cephalopod that you cannot provide the correct conditions for. For example, if your aquarium is a few thousand gallons in size with a powerful chiller and you have a healthy bank account then by all means get a giant Pacific octopus (GPO), *Enteroctopus dofleini*. This species comes with many challenges, so unless you can provide for them they are best left to the public aquariums. A GPO would be far out of the reach of the majority of prospective cephalopod owners.

Some species should be avoided because of their toxicity. They might not be particularly aggressive, but they can easily kill a person. For this reason alone species within the genus *Hapalochlaena*, commonly called blue rings, should be avoided. Blue rings are completely unsuitable for a home aquarium, despite the fact that individuals are often offered for sale. But be warned: *all* octopuses are venomous, and a bite from any species can and will hurt, even if not as much as a bite from *Hapalochlaena*.

Some species about which very little is known about their natural population status could possibly be

Old Octopuses

One of the hardest scenarios to avoid is buying an old specimen. That is difficult to avoid, because many times the octopuses being sold have not been correctly identified and you could be looking at a baby giant or a geriatric dwarf. Buying an adult will result in a very short-lived experience with it—the animal may have only a few weeks left to live.

threatened in the wild. This would include the zebra octopuses, which group includes species like the mimic, wunderpus, and *O. chierchiae*. Purchase of these species should be avoided to prevent their harvest for the pet trade.

Some species have a very poor survival rate from their shipping point and should be left untouched in the sea for this reason. Many of these specimens exported from their country of origin are either dead by the time they arrive at their destination or die shortly thereafter. We as conscientious hobbyists can make a difference by not creating a demand for such animals. Species with a very poor survival rate include the mimics, wunderpuses, and the deadly blue rings.

WHERE TO BUY YOUR CEPHALOPOD

At present there are a few different sources from which to obtain a cephalopod.

Aquarium Stores

In general, aquarium stores will not stock octopuses or cuttlefishes. They will need to be ordered for you. Before you place an order, see whether the seller will tell you which species they can obtain and whether they know what part of the world each available cephalopod comes from. Knowing the location goes a long way towards getting an accurate identification. It is best to arrange to pick up the animal when it first arrives at the store, reducing the stress placed on it by eliminating a second acclimation process it would have had to endure. Try and be there for the delivery and take your fascinating new purchase home in its original bag.

If the store has cephalopods in stock and on display, they will probably be housed in a small plastic tank within a larger tank. If the animal you're interested in is hiding away, ask whether you can see it so that you can observe it feeding. Most reputable dealers will gladly comply. If it is hiding inside a pipe or a shell, it would be easier and better to be given (or to buy) the hiding place too.

Try not to buy any species on impulse. For example, you might go to a store to pick up a species such as *O. briareus* and find that they have a mimic. It might be very tempting to bring the mimic home anyway, but this is a much harder species to keep and potentially rare in the wild. Also, most mimics travel badly and often die soon after importation.

Ask as many questions as you can about the octopus. How long has it been in the store? Has it been eating okay? What has it been eating? Has it been active during the day? What are its water parameters now? And so on.

The Internet

There are now many online aquatic stores that will deliver your purchase straight to your door, and many of them stock cephalopods. Before ordering one, ask all of the questions you would at a store if you were there in person. Also, more importantly, ask to see a picture of the actual animal. Some online shops use stock photographs and use the same picture no matter what species they are selling. We learned of one instance in which the pictures used in an advertisement were of a deepwater Antarctic species of octopus!

Remember that by ordering online you will have the extra expense of freight, which in some cases can double the total cost of your order. It is important that you specify overnight shipping and pay extra for early arrival, if possible. You will do better ordering by phone, where you can discuss the shipment, the weather, and other factors with the provider. Lastly, for your own safety check to see whether your order is refundable if the box goes missing or if the animal is dead on arrival.

Licensed Collectors and Other Sources

Licensed collectors can be found on the Internet and will often sell to the general public. They are educated in proper collection techniques, and their animals will be less stressed than those that have passed through multiple holding systems.

You might also be able to find your octopus through a fellow hobbyist. Also, depending on where you live and the local laws, you might be in the lucky position of being able to catch your very own cephalopod. Quite a few people have successfully caught their own specimens and got them to adapt to living in an aquarium. Some people have even managed to buy living cephalopods from fish markets selling cuttlefishes

A young *Octopus briareus.*

destined for the dinner plate! Scout around your local fish market or harbor area and talk to local fishermen. Hey, you never know...

TRANSPORTING AN OCTOPUS

Getting your octopus home from the aquarium shop should be similar to the way in which you might get a fish. The one major difference is that an octopus is a much harder beast to catch and get into a bag. The ideal scenario is to take a large bucket with a secure lid and siphon water from the shop tank into the bucket, then move the octopus, complete with hiding place, quickly into the bucket. This works well with small plastic tanks too. At home it is always much easier to acclimate an octopus in a bucket.

If you are going to have a long journey from the store to your home, prepare the bucket by drilling a hole in the lid just wide enough to slide an airline pipe through. Use a battery-operated air pump to provide some oxygenation. In most cases, though, this won't be necessary.

Keep the bucket or bag warm by wrapping it in a blanket. Alternatively, on a hot day you may need to worry about keeping the bucket cool, so always go straight home after picking up your new octopus. Don't stop halfway to do the grocery shopping!

If you order an octopus from an online shop the seller will often pack the octopus in two or three strong bags, filling the airspace with oxygen and placing the bag in such a way that it doesn't bounce around within a polystyrene box that may or may not have a cardboard outer sleeve. If sending tropical species the vendor will normally supply a heat pack for the journey. The heat packs normally last between 24 and 72 hours and will prevent the animal from getting too cold. The heat pack should never touch the bag; it is often taped to the inside lid of the box and separated from the bag by newspaper. Cold packs are available for hot summer weather, too. Some aquarium shops may also pack the animal for you in exactly the same way. Just ask. (To learn the best way to acclimate your cephalopod once you get it home, see Chapter 5.)

BUT I WANTED A CUTTLEFISH...

Cuttlefishes are almost as popular as octopuses these days, and that is primarily because of the availability of the eggs of a dwarf species from Bali, *S. bandensis*. Not long ago this species was all but impossible for US-based hobbyists to obtain. The unavailability of cuttlefishes in general resulted partly from the fact that there are no cuttlefishes found naturally around the shores of North America. (European cuttlefish keepers had a much easier time—there are several species in the East Atlantic.) Additionally, cuttlefishes were notorious for shipping poorly, and many died during transit.

On Shipping Cuttlefishes

If at all possible, always avoid shipping anything larger than a juvenile cuttlefish. Larger cuttlefishes do not travel very well in a bag. Getting cuttlefishes sent as eggs is an even better (and less expensive) option.

Pros and Cons of Owning a Cuttlefish

Keeping a cuttlefish is quite a different experience from keeping an octopus. It too has its pros and cons.

Pros

- Unlike octopuses, cuttlefishes cannot climb out of a tank.
- For the most part they can be kept socially.
- You can utilize an aquarium cleanup crew, because cuttlefishes will not bother with hermit crabs, brittle stars, and most snails after realizing their defenses are too much hard work to break through.
- Cuttlefishes are generally active for much longer through the day when compared to octopuses, therefore giving you more chances to watch them.
- They will not rearrange your aquascaping.
- The commonly available *S. bandensis* is quite often captive bred, thereby taking pressure off wild collection.
- They can be bred by most people who keep a small group.
- The freight charges for eggs are much cheaper.

Cons

- They don't always travel very well.
- They are not be able to change shape and are less likely to explore toys within the aquarium.
- Young cuttlefishes are very particular about eating live food only—and lots of it!
- They can be prone to dashing off backwards into the walls of the aquarium and badly hurting themselves.
- Many species need very large tanks and systems.
- Cuttlefishes are much more prone to inking than octopuses are.

Another traditional drawback with keeping cuttlefishes was that the species available was usually *S. officinalis*, which could grow too large for the typical home aquarium—about 18 inches (46 cm). They require a 200-gallon (800-liter) tank at the very least.

The real benefit with *S. bandensis* is that individuals are fully grown at only a few inches and can therefore be easily kept as a colony. Captive breeding is fairly easy. Rearing the young is a bit more problematic, as they require a lot of live foods, but captive-bred *S. bandensis* eggs are becoming fairly easy to find. They appear on the trade lists of many shops, although sometimes misleadingly advertised as squid eggs. If you are unsure, ask to see pictures—cuttlefish eggs will look like a bunch of black grapes.

Catching a cuttlefish for the trip home is normally harder than catching an octopus. You can normally rely on the octopus's hiding in something that can then be lifted out of the tank. If a cuttlefish is lifted out of the water it will very frequently ink when returned to water (i.e., the transport bag or bucket). Although nontoxic, the ink can coat gills and cause asphyxiation. Transferring a small cuttlefish is best done by encouraging it into a glass or jug by guidance with a net and lifting the glass out of the tank. Larger

Cuttlefishes, like this *Sepia bandensis*, are growing in popularity in the U.S.

Sepia bandensis.

cuttlefishes can be more problematic—one of the easier ways is to guide the cuttlefish into a large plastic bag underwater. If it is all done smoothly and with no sudden movements, you should be able to avoid an inking session.

One of the authors of this book, Colin, found out the hard way about cuttlefish inking with a 6-inch (15-cm) *S. officinalis.* It turned the water of several buckets jet black with ink before finally "going quietly." Subsequent captures using the previously described methods have worked well with cuttlefishes up to about 1 foot (30 cm) in length.

KEEPING CEPHALOPODS: A WONDERFUL EXPERIENCE

Cephalopods are not for everyone, but if you are up to the challenge, then read on. In the following chapters we'll show you how to set up the best tank for your octopuses and cuttlefishes and how to enjoy and interact with your new aquatic pets. And if you are up for a really exciting challenge, you might possibly encourage some of these captivating animals to breed in home aquariums.

Most importantly, this niche of the aquarium hobby is still in its infancy. There is much to learn, so much so that anyone who keeps an octopus or cuttlefish has the chance to discover something new about these breathtaking and fascinating animals.

Chapter 3

Tanks, Equipment, and Water: Planning the Octopus or Cuttlefish Aquarium

Octopuses are famous for being able to squeeze through tight gaps, slide under rocks, and cram themselves into impossibly small-looking holes. They have been known to solve simple puzzles and unscrew jars for the reward of food. Cuttlefishes are well known for being able to change color and skin texture in a bewildering display. Both octopuses and cuttlefishes can respond to their owner's presence and often get cited as displaying alien-like intelligence. One of the reasons you probably want to keep

a cephalopod in a the first place is to witness all this amazing behavior. In order to experience this, you need to set up an artificial habitat suitable for the cephalopod to show off its amazing life skills. You need to design a mentally challenging underwater assault course rather than just setting up a run-of-the-mill normal home aquarium.

THE BIGGER THE BETTER

The most important issue to consider before you buy a cephalopod is that the size of the aquarium has to be proportional to the potential adult size of the animal that it is intended for. You should get a tank of that size at the very beginning.

Put simply: make sure you find out what species you are planning to keep and that you buy a big enough aquarium to house your cephalopod for its entire lifespan. Being moved from tank to tank does not suit any octopus or cuttlefish. Tank size is something that cannot be skimped on and requires a lot of research relevant to your intended species.

To get an idea of how large your aquarium should be, we give some suggested size guidelines in Chapter 8. These aquarium dimensions are suggestions based upon hobbyists' experiences of keeping them in a normal home aquarium. To summarize those guidelines—and you might get tired of hearing this—but bigger is always better!

In a captive situation we should always strive for the best possible environment for our cephalopods, and ultimately the water quality of the aquarium is tantamount to success. Water parameters will be discussed later on in this chapter, but one of the important points to be discussed here is: the larger the aquarium, the larger the body of water, and therefore the more stable the aquarium will be for your cephalopod.

As an example, a small aquarium of less than 30 gallons (120 liters) can heat up or cool down quickly if the room's ambient temperature increases or decreases. Also, an item of uneaten food can quickly rot and pollute a small tank. By comparison, a larger tank of 100 gallons (400 liters) or more will fluctuate much less and allow a much longer period of time for the owner to detect a problem.

If there is some uncertainty in regard to what species of octopus or cuttlefish you are obtaining, it's better to err on the side of caution and get as large a tank as you can. It is suggested that a 30- to 50-gallon (120- to 200-liter) tank is the minimum for most of the commonly seen species, but be warned that some of the species easily available to home aquarists should be housed in nothing short of a 200-gallon (800-liter) tank.

Variations in Size

You should be aware that within a species there is a certain amount of variation in size just as in humans. Some specimens are able to reach a substantially larger size than their conspecifics. There may be genetic reasons behind this disparity between individuals, but there is no doubt that diet and husbandry play a large part in determining the eventual adult size.

A TANK WITHIN A TANK

This technique is often utilized in aquarium stores and can be an easy way of maintaining a very small octopus for its initial introduction to aquarium life. It simply involves a small plastic tank with a tight-fitting perforated and escape-proof lid situated within a larger mature aquarium. There are quite a few manufacturers of these types of aquariums and a variety of styles, but all follow the same basic theme.

An octopus in the smaller tank cannot get into the rest of the aquarium. This is how many aquarium shops keep their octopuses until they are sold. Although it's normally only shops that house octopuses this way, the method can also be useful for housing dwarfs and immature individuals at home, as it's a good way to quarantine or acclimate the octopus. The timeframe for this could last a few days or a couple of weeks, but it provides a great opportunity to make sure that the animal settles into captive life and that it is eating well.

Keeping your cephalopod in a small container inside your main tank will help it adjust to your tank without being too stressed.

We must stress that the larger aquarium—in which the smaller aquarium is housed—should be fully matured (have a working biofilter) and should have good water flow to ensure that there's water exchange between the two tanks. There should be something in which the octopus can hide within the smaller tank—it should never be bare. Also, don't allow the small tank to tumble about in the water column; weigh it down with something.

A new octopus could be kept this way in the home aquarium if it is very small and likely to get lost within the large tank. It will also make feeding the small octopus much simpler. Placing food in the small tank will make it easy for the octopus to find it, and you will know exactly how much is being eaten, or not eaten as the case may be. Always remove uneaten food after a few hours.

Please note that this is just a short-term housing solution. The ideal octopus tank has been set up specifically and only for octopuses.

THE AQUARIUM

When you are ready to obtain an aquarium, it is often best if you go out and visit several different pet shops and look at the off-the-shelf fish tanks they carry in stock (ones not part of a starter kit). Most aquatics shops will have a selection of tanks in all different shapes and sizes. Browse the Internet, looking at the various aquarium manufacturers' designs to see what catches your eye. Look for something that will be practical and pleasant in your home. Bear in mind that you will need to make the tank escape proof.

Better Safe Than Sorry

Cephalopods are very sensitive to copper and will die if exposed to even small amounts of it in the aquarium, so it is always safer to buy a brand-new tank rather than a secondhand one, unless you can be 100-percent certain that no copper-based medications have been used to treat past inhabitants. The same goes for converting any freshwater tanks to one for your cephalopod.

It will be easier to do this to some models than others because this is not an issue that manufacturers consider when designing their systems.

Aquariums are normally made from glass or acrylic plastic. They come in many standard sizes, including rectangular oblongs, cubes, cylinders, pentagons, hexagons, L shapes, and all manner of custom designs. They may be tall and narrow or short and wide; they may be sealed with clear silicone or black silicone and have open tops, built-in filters, overflows, drilled holes; you could even get two tanks joined together with clear tubes. Manufacturers have their own names for all the variations on the same basic theme: a waterproof enclosure created to display aquatic animals. Aim for a long-and-wide shape with as much surface area as possible.

The aquarium you buy will become your cephalopod's entire universe, and that includes being its life-support system. If things go wrong, the animal can quite simply die. But if all goes well the cephalopod keeper will be rewarded by seeing the animal live, grow, and possibly even reproduce.

Glass Versus Acrylic

Why choose one over the other? Each material has its advantages and disadvantages, and at the end of the day the choice will just come down to a personal preference. However, in order to help you come to a conclusion, visit a few different aquarium stores and seek advice about which types of tanks are available near you. The choice of material that the aquarium is made of makes no difference to the cephalopod as long as it is safe for marine aquarium use.

Shape of the Tank

The shape of the tank should be given careful consideration in regard to the proposed inhabitants. For the most part a cephalopod aquarium should ideally be long and wide rather than tall and thin. Cuttlefishes and octopuses follow a mainly benthic existence, and the larger the base surface area the better. There are no hard and fast ratios for length versus width, but you should favor width over height in tank dimensions in almost every instance.

The captive husbandry of squids and nautiluses requires a different approach in their aquarium's requirements. A nautilus, which in the wild goes through a daily vertical migration, would do better in a tall tank. Squids, which tend to be very difficult to keep long term in any type of aquarium, tend to do slightly better in a circular tank or a tank with sloping edges, as this lessens chances of their crashing into a solid wall while swimming quickly backwards, which can cause injuries, infections, and death. These specialist tanks will be discussed in more detail later in this chapter and Chapter 8.

A type of aquarium that has become increasingly available in the last few years to a more general market is called a kreisel tank. It is often used for aquatic animals that require being in constant motion. These tanks are most commonly used in public aquariums to display invertebrates like jellyfish. They are circular in design and use water pumps to keep the water column in constant motion. Perhaps a design based on this principle is the answer for the display of squids, but at the present time there is no ideal solution for housing squids other than being maintained in fiberglass tubs with sloping edges. They do not do well in the captive environment and should be avoided and left for public aquaria or research institutes. Only one squid group may be suitable for captive care, benthic species such as the bobtail squid. However, at the time of writing these squids are few and far between in the aquarium trade.

Placing the Tank

Where do you want to keep your pet cephalopod? It's actually a more complicated question than you might initially think. Obviously you want it situated in an area where you will be able to sit, relax, and watch your cephalopod. You maybe also want it in a place where the whole family can watch and learn. Perhaps you are lucky and have a whole room set aside for animals? Whatever the location, there are some do's and don'ts when it comes to positioning cephalopod aquaria.

- Make sure that the floor the tank and aquarium stand will sit on is strong enough to support the

Your tank must be set up and fully cycled before you add your new octopus or cuttlefish. This is *Abdopus aculeatus* in a typical pose, with eye horns.

weight. A full tank is very heavy—figure about 10 pounds per gallon (1.2 kg per liter)—and should be positioned against an outside wall running along as many floor joists as possible (where applicable).

- Ensure that there are enough electrical outlets nearby. The cephalopod aquarium normally has more than a few items of equipment that require their own electrical supply. Be sure these are all GFI outlets; this is for your own protection.
- Do not position the aquarium next to a heat source, such as a radiator, a fireplace, or a window, as this can overheat an aquarium. Likewise it makes sense to keep the aquarium away from air conditioning units, as this will chill the water.
- Position the aquarium away from sources of loud noise. A slamming door or a bass-thumping speaker from a stereo is enough to frighten a cephalopod, causing it to ink in the aquarium! Although they don't have ears as we understand them, they can pick up low sounds.
- The aquarium shouldn't be situated in areas of heavy traffic such as a hallway because the animals will frequently get spooked and ink if people suddenly appear. The thundering footsteps of children running wild are best avoided.
- If you are planning to decorate the room in which your tank will be situated, make sure you do it

It's a good idea to have a setup in which you can store all of your equipment underneath the tank. This consolidates everything you need for your cephalopod into one neat and out-of-the-way place.

before the tank is placed. Aquariums are very difficult to decorate around, and once again it is a disturbance best avoided for a cephalopod. In addition, fumes from some paints and solvents can be poisonous.

Costs

How much you spend on a tank is ultimately dependent on what you can afford to spend, but it is fair to say that keeping a cephalopod is not something that you can do on a shoestring budget. There are things you can do to keep prices down, such as not blowing half your funds on the stand because it has to match the rest of your living room's furniture, but you should not skimp on the aquarium's equipment budget. (What equipment you need for a cephalopod aquarium will be discussed later on in this chapter.)

Before you part with any money make sure that you visit a few aquarium stores and shop around. Look for a deal that suits keeping a cephalopod. Know the equipment you need and the size of the aquarium you require. Try not to go for the first nice aquarium you see; compare those of different manufacturers.

THE WATER IN THE TANK

There are no freshwater cephalopods. Some species inhabit tidal zones or estuaries where the amount of salt in the water might fluctuate, but for the most part cephalopod husbandry requires that the salinity of the water is as close to that of true sea water as possible.

Salinity

The salinity of water is typically between 34 and 35, meaning 34 and 35 parts dissolved salt per thousand. This is *usually* what is needed for a cephalopod aquarium, but is not absolute. Research the habitat of the species you are planning to keep and adjust the salinity accordingly before you get it.

When it comes to reading the salinity of an aquarium there are two things to bear in mind: salinity changes depending on the water's temperature, and salinity cannot be measured directly. The most commonly used device to determine how much salt is in a marine aquarium is a swing-arm hydrometer. These work by measuring the density of water. Pure water has a density of 1 and as far as we cephalopod keepers are concerned we should aim for a density range of 1.024 to 1.026.

Swing-arm hydrometers can be purchased from aquatic shops and are fairly accurate, but they need to be cleaned between tests, as salt can jam the mechanism and make them read incorrectly. Being dropped or jostled can also throw the reading off. A backup hydrometer is good to have on hand—you can use it take a second salinity reading and compare it to your first hydrometer's reading. Floating glass hydrometers are also used, but they are very fragile.

Other methods are being seen in modern home aquaria, including the use of electronic salinity monitors and refractometers, which were once much more expensive and restricted to laboratory use only. The electronic meters measure the conductivity of the water (ions dissolved in the water) and express it as an amount measured in millisiemens (mS). They need occasional calibration but should come with necessary standards.

A refractometer is a very easy item of equipment to use and is calibrated with distilled water. By looking

Don't Play With Your Specific Gravity

Many aquarists who keep marine fishes will often keep the salinity a bit lower than natural sea water, as it can help combat parasites. This is not recommended for cephalopods. They always do better in full-strength sea water. This may also be something to consider if you buy salt water from an aquarium store. You may have to ask for the appropriate specific gravity or be prepared to add additional salt before using the water.

through the refractometer towards a bright light with a sample of water in the slide, you can clearly read the salinity on a scale that normally includes specific gravity (SG) and a conversion into parts per thousand. Always refer to the manufacturer's instructions when using a refractometer.

By far the most important aspect of an aquarium's salinity is that a constant balance should be maintained. Always keep the salinity at the same level and replace water lost to evaporation with pure reverse osmosis/deionized (RO/DI) water. It is much better to replace a small amount each day rather than large amounts infrequently.

How to Make Salt Water for the Cephalopod Aquarium

When you first set up your tank and before anything is added, first clean it—using water from your tap is fine. This will remove any dust and residue that may have collected during the tank's manufacture and subsequent storage in a store. Do *not* use any cleaning products or chemicals; a rinse with tap water will be fine.

Before you add any living animal or any rocks or sand harboring life to the tank, you must fill it with salt water. Ideally this should be pure RO/DI

Constantly tampering with your water's chemistry will cause undue stress to your cephalopod.

water mixed with a high-quality aquarium salt. This pure water can either be bought from your local fish store or come from your own water purification system.

Always give your aquarium the best possible start—it is not as simple as just pouring in municipal tap water. Fill the aquarium with RO/DI water at the correct temperature and add salt. Adjust accordingly until it gives a reading of between 1.024 and 1.026. Experience will make this a faster process. Try counting how many cups to a bucket, for example.

Ideal Water Parameters For Cephalopods

- Salinity: 1.024 to 1.026
- pH: 8.2 to 8.4
- Temperature: depends on species, see Chapter 8
- Ammonia: 0
- Nitrite: 0
- Nitrate: <25ppm
- Regular water changes will keep trace elements in good supply.

Measuring Water Parameters

Each and every aquarium, whether it contains fresh or salt water, should have its water tested on a regular basis. Testing aquarium water regularly should be an early warning device, allowing you to anticipate any problems that might be developing. It is basically akin to keeping your, as the saying goes, finger on the pulse. An aquarium may have sparkling clear water at a glance but it can be a toxic bath to anything living in it.

Correcting the SG and pH

If for any reason your regular tests show that the conditions within the aquarium have deteriorated, you must correct this situation immediately. The SG and pH have more than likely changed gradually over some time in the aquarium, and therefore any changes you make to rectify the problem should also be done gradually.

Through evaporation the salinity will rise each day. This increase in salinity should be balanced by adding back the appropriate amount of RO/DI water. Evaporation happens in any aquarium, and one of the best ways to keep this in check is to draw a line on the sump tank at the perfect SG and add water when the level drops below that line. (Sump tanks are discussed later in this chapter.)

Adding a pH and/or KH buffer will combat the aquarium's tendency to gradually drop in pH. It is perfectly natural for this to happen and will be evident in any aquarium. If you use an aragonite substrate, it will help stabilize the pH. There are many products on the market to buffer aquaria and they are very similar in nature. Just make sure that any adjustments to pH are done gradually and not in one fell swoop. Follow manufacturers' instructions carefully.

Ammonia, Nitrite, and Nitrate

By using standard liquid test kits you will be able to monitor ammonia, nitrite, and nitrate levels within your tank. In an ideal world all three of these tests would show a zero result, but this is not an ideal world.

Ammonia is the most toxic of the three and is the most toxic part of the nitrogen cycle within our aquariums. It needs to be avoided at all costs. Before you combat the ammonia you need to find out what caused it. Look for uneaten food, dead animals, or a biological filter that isn't working. You can combat ammonia by carrying out a large water change and possibly adding new filter media for adsorbing ammonia.

Octopuses produce large amounts of ammonia. You must have a working biofilter before your cephalopod is added to your tank.

Nitrite is not as toxic as ammonia but is still dangerous, and is the next stage in the nitrogen cycle. If you have an unexpected spike in nitrite, you should test the water for ammonia (in case you haven't already) and look for a possible source. A nitrite spike can be caused by a faulty filter and having the nitrification cycle starting to work in reverse. Water changes carried out a few days in a row will help to combat this scenario.

Nitrate is the least toxic part of the nitrogen cycle. Some of it will be converted by bacteria into harmless nitrogen and will leave the aquarium through the water's surface. It is not harmful in concentrations up to 50 parts per million (ppm), but from there onwards it can have a detrimental effect on marine animals, especially breeding animals. Again, regular water changes should keep your nitrate levels in check.

Nitrate filters are commonly seen for sale these days, and they are normally quite effective. You should also look into having a sump converted into a refugium with macroalgae and detritivorous animals. A refugium set up with a light cycle on at the opposite time of the main aquarium can also help prevent daily pH swings.

The Nitrogen Cycle Within an Aquarium

As an experienced aquarist you should already have a good understanding of what happens during the nitrogen cycle in regard to filtration and general aquaria. If you require a refresher on the subject, you should refer to any number of available books that describe in more detail the chemical reactions and changes of waste produced in the aquarium from ammonia to nitrite and then nitrate. We will provide an easy way to mature (cycle) your new cephalopod aquarium in the next chapter.

Nitrate To Nitrogen

It is possible to complete the nitrogen cycle by utilizing anaerobic bacteria to convert nitrate to nitrogen, but more often than not good old water changes are the easiest way to eliminate nitrate.

NECESSARY EQUIPMENT

A cephalopod aquarium functions using all of the same equipment that a normal marine aquarium uses. There are some slight differences to make it suitable for octopuses, but in essence there are more similarities than differences. The aquarium requires a form of heating (or cooling), filtration, protein skimming, lighting, and test kits to help maintain the water parameters. Around this essential equipment there are other items that not only help maintain the aquarium but also actually make maintenance much easier for you. But buying all of the equipment is the expensive part of the hobby.

Your Cephalopod's Life Support System

The equipment of a cephalopod aquarium is best thought of as a life support system. It is there to keep the animal alive, and if it is not suitable for the job the octopus or cuttlefish will die.

The octopus aquarium relies on a good deal of forethought and planning when it comes to equipment and tank setup. All of the equipment literally has to be glued down or kept well out of the reach of inquisitive arms and suckers. An octopus will decorate and redecorate its aquarium any way it wants to. It will push and pull rocks about, dig in the substrate, and make caves, caverns, crevices, canyons, grottoes, and any other type of shelter it fancies; any item of equipment, even heaters, will be fair game for such activities—and possibly broken with dire consequences for animal and keeper alike. Broken aquarium heaters have been known to cause house fires!

By far the best method of achieving a safe-and-effective filtration system is to incorporate a sump into your tank's system and have an overflow method of filtration.

In modern aquaria it is now quite commonplace to be able to buy fish tanks that are either pre-drilled or supplied readymade with an overflow and complete external filter setup. In such a system, water leaves the aquarium via a hole drilled through the glass and goes into a sump tank via pipes. The sump tank is normally situated in a cabinet under the aquarium. The filtration system, protein skimmer, heater/chiller, and all other pieces of equipment are situated there as well, including a water pump that returns the water to the aquarium.

Ideally the overflow holes are behind a weir that prevents the octopus from escaping through the hole into the sump and prevents it from using its body to block the hole and creating a flood as the water level rises in the main aquarium. The surface film of water within an aquarium is where most of the proteins and other waste products are found. It is always better to have the water leaving the aquarium leave from the surface so as to remove the buildup of waste. You can see the surface film from looking through the front glass and up at the surface. It should be clear and not dusty gray.

A cuttlefish aquarium doesn't need to be quite so strictly robust, but it is useful and more aesthetically pleasing to have the bulky equipment out of the tank and in the sump.

Within the sump you can house the equipment discussed in the following sections.

Filters

A filter does much more than just keep the water clean. The filter is one of the most important parts of the aquarium's life support machine and needs to be the right type for the job. There are many different types of filter, and in fact it could be argued that an aquarium set up for cephalopods would be all the better for taking advantage of these different methods of filtration simultaneously. What's even more complicated is that there are many manufacturers with slightly different angles on the same basic structure and theory. In

No Undergravel Filters

Undergravel filters will rarely succeed in a cephalopod aquarium; octopuses (and some cuttlefishes) will dig into the substrate too much to let an undergravel filter be effective.

any case, there are three types of filtration that you should be cognizant of.

Mechanical Filtration

Mechanical filtration is the physical removal of waste from the water column. This type is perhaps the method that people first think of when it comes to filtration. In its most basic form, the water passes through a fine medium or media that catch suspended particulates and prevent them from going back into the aquarium. There are different grades of filter material that will catch different sizes of particulate, and a mechanical filter should be set up to the specification of your tank. The two most popular types of media are sponges and filter "wool." Both of these are specially made for the aquarium industry and will be nontoxic for marine aquarium use. The use of other sponges not meant for tanks may leach poisonous chemicals into the water. Another medium is a micron bag that has very fine mesh and will catch most detritus.

The main advantage of a stand-alone mechanical filter—and where that filter really comes into its own in a cephalopod's tank—is in its ability to capture the shed skin from an octopus or cuttlefish's suckers. A mechanical filter is also an essential item to have if you plan to have a sump as part of your aquarium.

Mechanical filters are best placed at the first step of filtration: where the water leaves the aquarium. If the mechanical filter is part of a larger filter and you are using a sump, the water leaving the tank should be passed through a mechanical filter first.

Mechanical filters should be cleaned on a very frequent basis—possibly even daily in the case of a large octopus. Taking good care of your mechanical filter as part of a sump setup will give you a good pre-filter that will remove a lot of the solid waste before it gets to the protein skimmer or biological filter.

Chemical Filtration

The most commonly used chemical filter in the aquatic hobby is carbon. Over time the water of any aquarium will start to take on a slight yellowish appearance—this is a buildup of organic chemicals within the aquarium. Water changes will combat this, but adding a carbon filter to your sump will keep the water sparkling clear. The carbon adsorbs chemicals, binding them and removing them from the water column. For this reason the carbon should be changed on a regular basis (usually monthly) to prevent it from becoming exhausted. If an octopus or cuttlefish has a heavy inking session, replace the carbon after a couple of days—the ink will have been adsorbed into the carbon.

Other chemical filters on the market include pillows specially made for aquarium use that remove nitrates and phosphates. These can be a real benefit to any aquarium, as they can help prevent buildup of the nutrients that encourage algal blooms.

Biological Filtration

A biological filter utilizes beneficial bacteria to reduce harmful organic chemicals in the water to less harmful chemicals. Beneficial aerobic bacteria live on every surface of the aquarium, from the inside panes of the glass to the surface of the substrate, but to encourage sufficient numbers a biofilter contains a medium with a vast surface area and porosity at the microscopic level. This provides ample sites for the growth of beneficial bacterial colonies. There are many species of bacteria that provide biological filtration, and the best thing we can do is provide optimal conditions and substrates for the bacteria to live and multiply on, especially a steady flow of highly oxygenated water through the biomedium.

Aerators

Octopuses and cuttlefishes have a high demand for oxygen within the aquarium, requiring a steady supply of well-oxygenated water.

Because of the very nature of a biological filter and the fact that its bacteria are aerobic—meaning that they need a good source of oxygen—water returning to the main aquarium might be oxygen poor, since the bacteria will have stripped the water of this vital gas. Therefore it is always a good idea to heavily aerate the water at the end of the sump just before it returns to the aquarium. Do not use a bubbler to aerate the water—octopuses especially may sit on the bubbles and get air into their mantles, which could lead to problems. You might be able to use an air lift, but there's a good chance your octopus will turn this into a toy, so avoid adding air directly into the aquarium. Effective aeration can often be provided by arranging the return so that the water returns to the aquarium above the surface of the water. Protein skimmers are also excellent providers of gas exchange.

Protein Skimmers

Many an octopus and cuttlefish owes its life to a protein skimmer. This piece of equipment really is a life saver. It is adept at removing not only proteins and other organics from the water on a daily basis but also ink in emergencies. If an octopus or cuttlefish inks within the aquarium, the resulting pollution can be so bad that it is literally impossible to see into the tank. In such cases the concentration of ink can lead to asphyxiation and death if not taken care of. It is always a good idea to immediately perform at least a 25-percent water change in the event of a heavy inking. In the meantime the protein skimmer will work to remove the ink. When this happens the collection cup of the skimmer will need emptying sooner than usual.

Various filter media are available, like these plastic shapes, and all are designed to provide good homes for beneficial bacteria to grow.

For this reason the protein skimmer is not an item of equipment that you should leave off your

A protein skimmer is an essential piece of equipment for the cephalopod aquarium.

shopping list or skimp on. The bigger the better, and because of the amount of waste produced by the cephalopod, its messy feeding habits, and the odd inking session, a skimmer is worth its weight in gold! Aim for a skimmer rated for an aquarium about 50 percent larger than yours. They can be expensive, but the first time your cephalopod's life is saved by a protein skimmer you will be glad you spent the money.

There are a few variations of the protein skimmer, and they can be housed in different parts of the aquarium. Some are internal, some are external, and some hang on the back of the aquarium. A protein skimmer situated inside the main aquarium will just be seen as a toy by most cephalopods and will never function correctly. Worse still, all the waste it has collected might be dumped back into the tank! The hang-on-the-back protein skimmers, although mostly external, might offer an escape route for an octopus or a possible means of flooding if the octopus can jam a pipe inlet or outlet. The best are the ones designed for use within the sump.

Each potential skimmer should be evaluated carefully for its own merits. If in doubt, ask about the particular brand with someone who has kept an octopus, and specifically how they octo-proofed the equipment.

Heating/Cooling Devices

This next piece of equipment is entirely species-dependent. If you are in a temperate part of the world and are planning to keep a tropical cephalopod you will need a heater. Conversely, if you live in a tropical climate and want to keep a temperate species you will need a chiller. Even tropical species may need a chiller in summer if you experience extremely hot weather.

There are many different aquarium heaters, and they almost always incorporate their own thermostat. You should compare the capacity of your aquarium (and sump!) with the correct wattage of heater as per the manufacturer's instructions. Normally it is better to have two 150-watt heaters as opposed to one 300-watt heater. In case one fails you still have the other heating your tank. Set the aquarium heaters to the desired temperature suitable for the species you are planning to keep. As with the protein skimmer and other pieces of equipment that if kept within an aquarium domiciling a curious and downright manipulative dweller like your average octopus would be subject to being pushed, pulled, dragged, carried, hauled, toted, shoved, and deposited all over the tank, it's best to situate the heater in a sump, not the tank proper.

Heating an aquarium is altogether easier than having to cool an aquarium. There are several different ways in which this can be achieved. By far the easiest, although perhaps the most expensive, is to buy an aquarium chiller.

Lighting

Few cephalopods will appreciate bright lights. The intense metal halides that are so popular with reef tanks don't belong in many of the aquariums used for an octopus or cuttlefish; they are simply too bright.

Most species prefer a tank which is lit by standard aquarium lights such as T8s. It is less harsh on their eyes and, as many cephalopods are either crepuscular (primarily active at dawn and dusk) or nocturnal (active at night)—with only a few being diurnal (active during the day)—they are more likely to be active in less harshly lit tanks. Likewise, many people also believe that the blue from actinic bulbs is irritating to a cephalopod's eyes.

There is a huge range of aquarium lights available to the modern hobbyist, and any of the bulbs for aquariums with a color temperature of 10,000 to 15,000 K will be fine.

For viewing shy individuals or nocturnal species you could try using red lights, which are invisible to cephalopods that would otherwise hide from normal aquarium lights. You could try normal aquarium lights through the day with the option of switching on a red light for nighttime viewing when you switch off all ambient light in the room.

In some situations you might find that the ambient light in the room is enough for your cephalopod

If you decide to keep corals with your cephalopod, you must make sure your lighting will sustain the corals but will not harm your cephalopod's eyes. Pictured is *Octopus bocki*, a dwarf species, photographed off Lizard Island, Australia.

aquarium and that you don't need any further lighting. In fact this is probably most applicable if you were setting up a captive breeding program and there were multiple tanks as part of a larger system. Aquarium lights for cephalopod tanks are really more for our benefit than the animals in them, unless of course you have a tank with some cephalopod-friendly invertebrates that require light, like nonstinging corals.

The last few years have seen LED lights becoming more popular. One of the benefits with these lights is that they give off very little heat compared to a normal aquarium light. Don't use blue LEDs to illuminate your aquarium at night; use red LEDs for nocturnal viewing.

The photoperiod of an aquarium should be maintained at a similar length to the cycle these animals would have in the wild. You can easily do this by setting the lights to come on and go off at the same time each day by using a timer. Twelve hours on and 12 hours off is good balance to aim for.

TANK MAINTENANCE

The life of your octopus or cuttlefish depends on your maintaining the aquarium and keeping it in tip-top condition. Some of the tasks must be carried out daily, some less frequently. To make sure you keep on top of it all you should maybe think about starting a diary, a log, or perhaps even a spreadsheet or database.

Daily Tasks

The following is what you should be doing on a daily basis to keep your cephalopod aquarium on the up-and-up.

Appearance

Check the appearance of your cephalopod. This may not always be possible—these guys are often very secretive—but get to know your animal. If it is normally active at feeding time but is hiding away the next time you feed, you should start looking for the cause of its timidity. Study the animal's breathing rate and its demeanor. If something seems wrong, find the cause and correct it.

Temperature

Is the temperature where it should be? If the tank has gotten too hot, the cephalopod might be stressed. It might have an increased breathing rate or might even be trying to escape from the tank. If a cephalopod is hanging about at the water's edge or trying to get out of the tank, you should first check the temperature. You will also need to check your other water parameters.

Coral-Reef Cephalopods

Some of the cephalopod species that have evolved on coral reefs may be less affected by bright lights. Some hobbyists report that cuttlefishes like *Sepia bandensis* seem unaffected, but nonreef species like *S. officinalis* definitely become shyer under bright lights and spend more time hiding. Octopuses are certainly more active in less bright conditions.

If you decide to keep a refugium, you can grow various marine macroalgae in it, like this *Chaetomorpha*.

Equipment

Have a good look at the aquarium's equipment to make sure it is all working. It has already been said that the aquarium is a cephalopod's life support system and therefore has to be in perfect working order. Does the protein skimmer collection cup need emptying? Is the filter working? Is the water circulation at a normal level? Give everything at least a glance and get used to what to specifically look for.

Food

Is there any uneaten food left in the tank? This will quickly rot under normal circumstances and pollute a tank if not taken out. Be sure to remove any shrimp or crab carapaces, as they can increase phosphate levels in the tank.

Did your cephalopod eat as normal today? It's not a disaster if an octopus or cuttlefish misses a day or two of food, but if your animal goes off its feed, check for uneaten food, temperature fluctuations, faulty equipment, or poor water conditions.

Weekly Tasks

This is when you should check the water parameters. Set aside at least 30 minutes to test the water for all its major components. The pH should be between 8.2 and 8.4. Check the nitrogenous buildup in the water. Ammonia is the most poisonous of the animal's wastes. You should always get a zero reading for ammonia. Likewise, nitrite should also read zero. Nitrate should be as low as possible, but never higher than 50 mg/l.

You need to measure the SG carefully; this could even be a daily task if you have high evaporation from the tank. The salinity of an aquarium can change quite quickly from evaporation. You might not notice it from one day to the next, but as water evaporates the aquarium becomes more saline and can get to dangerous levels if you don't top it off with fresh water.

Use a good-quality hydrometer or, even better, a refractometer, and aim for a reading of about 1.025. The larger the tank, the more water you will lose on a daily basis through evaporation. In Colin's experience it can be as much as 2 quarts (2 liters) of water per day on a 200-gallon (800-liter) cuttlefish tank with a large trickle filter. You could look into setting up an auto top-off system in which a monitor can measure water levels and switch on to provide fresh RO/DI water from a separate holding tank when need be.

Other weekly chores would include emptying the protein skimmer's collection cup, washing the pre-filters of a filter, washing any salt creep from the tank and sump, checking the overflows for blockages, and removing algae from the front glass. Inside the tank you should consider using a gravel vacuum to remove any solid waste.

Monthly Tasks

This is when you should carry out a water change, although it can be done more frequently than this if it is required. Many cephalopod keepers will change a portion of the water every two weeks.

Your weekly water tests will indicate whether you need to do a water change sooner, so pay attention to them. The average tank will be fine with a monthly 25-percent water change. Make sure you give the

On Algae

Due to the nature of a cephalopod aquarium and the waste produced by these predators you can expect algae. The first bout of algae, as usual, might occur during the cycling of the aquarium and will pass with time before you add your cephalopod. Thereafter any large growths of algae may have been caused by overfeeding, under-filtering, or perhaps a buildup of crustacean exoskeletons resulting in a phosphate spike.

Many aquariums housing large octopuses or cuttlefishes are filtered with wet-dry trickle filters. They often get called "nitrate factories," and they can cause algae problems. Regular water changes are the best weapon against nitrate buildup.

The addition of a macroalgae refugium may also help to keep the unwanted algae at bay.

Once your tank is up and running and you add your cephalopod, always maintain healthy parameters. Pictured is Octane, an *Octopus* cf. *hummelincki* clinging to the aquarium's glass.

substrate a good bit of disturbance before siphoning out any water. This way you will remove solid waste from the substrate as well as from the old water. You could also shoot some water over the surface of live rock and over other items of the aquascape using a turkey baster or a powerhead. This disturbs any dirt that is blocking the pores of the rock.

UP AND RUNNING

Remember: keep it simple, buy good equipment, don't cut corners, let the tank's biofilter fully mature, carry out regular water tests, and perform water changes. If you can manage all of this, all the rest should fall into place and you can keep a healthy cephalopod.

Setting Up the Nautilus or Squid Tank

Nautiluses and squids are two cephalopods that are available in the aquarium trade from time to time. Keeping them is extremely difficult and should only be attempted, if at all, by the most advanced hobbyists.

Nautiluses

Keeping a nautilus in a home aquarium is not easy. It requires dedication and some expenditure, because you will need more equipment than for keeping an octopus or cuttlefish. You will need to learn more about both the behavior and biology of these cephalopods if you attempt to keep them.

The biggest issue with keeping nautiluses is the water's temperature. Because they naturally live in deep water they need cool temperatures. Room temperature is usually too high for them. You will definitely need a chiller to keep them between 50° and 70°F (10° and 21°C). Sustained temperatures above 75°F (24°C) are fatal. This is often a problem with buying pet-store nautiluses: it can take a couple of days after high-temperature exposure for them to die.

In addition, it would be a good idea to provide a fluctuation between these two temperature extremes over a 24-hour period. This will simulate the temperature changes in the sea when the nautilus migrates up and down the water column. For these temperature changes, you'll need a controller or a timer in addition to a chiller.

Light is much less of an issue. Since nautiluses live in deep water, they need little or no light. One lab both authors visited kept the side of the tank covered with a black cloth unless the nautiluses were being viewed. Other keepers use actinic light or ambient light, but even ambient light may be too bright. Of course, this means that the tank cannot house most corals, because it will be too dark for their symbiotic algae to photosynthesize. It also means that nautiluses are not a pleasant choice for a home aquarium, aesthetically speaking: they need to stay in darkness most of time to avoid damage to their eye pigmentation, so they are useless for an attractive aquarium display at home.

Nautiluses don't make much of a show tank for visitors to a public aquarium or as a focal point in your house. In fact, they have been described by some as being "as dull as keeping large floating snails." But to those who know and love cephalopods, they are fascinating to see, even without many of the traits that make other cephalopods so attractive.

Like other cephalopods, nautiluses produce a large amount of waste and need oversized filters to keep the water in good condition. And just like octopuses, they require a filter that can handle a tank two or three times the capacity of the tank used, as well as a large protein skimmer. This is also one of the few cephalopods for which a UV filter is recommended because of the need to minimize any possible pathogens. Like other cephalopods, nautiluses are difficult to treat if they fall ill.

Special care must be taken in designing the tank because the nautilus can run into any obstacle and sustain

Nautilus belauensis.

an injury. The best method for aquascaping a home nautilus tank is to place live rock only in the lower part of the tank.

Squids

Several individuals of the Hawaiian bobtail squid, *Euprymna scolopes,* can be kept for a time in a 30-gallon (120-liter) tank with a sand bed several inches deep. Some bobtail keepers have reported that their "bobs" don't bury themselves in the sand but do have the sand cover them as they rest on it. Again, they are nocturnal, so use a red light to view them. Juveniles should be fed *Mysis* shrimps, while adults can eat live or dead shrimps or small fishes.

For more information on nautiluses and squids (and argonauts, a third rare cephalopod), see Chapter 8.

Chapter 4

Aquascaping and Ceph-Proofing

It is very likely that the main reason you are considering keeping a cephalopod is because you heard how interesting their behavior is. Most people know that an octopus can squeeze through gaps the size of an eye or that a cuttlefish can change its color many times faster than a chameleon. Perhaps you have heard the story about octopuses unscrewing jars to receive food and that got you interested, or possibly the tale of an octopus leaving one aquarium, crawling across the floor to reach a second tank and eating all the inhabitants before returning to its tank, leaving a fishless aquarium and a damp trail.

WHATEVER your reason, a cephalopod is an animal that possesses a natural curiosity about everything and anything it can lay its suckers on and will display endless hours of interesting and even amusing behavior if you can provide it with the right habitat through careful aquascaping.

We will be using the word "enrichment" frequently in regard to providing a cephalopod with everything it needs to lead a fulfilling life. In its most basic form, enrichment is all about avoiding animal boredom. As you will see from some of the examples in this chapter, an octopus aquarium, unlike a reef aquarium, can end up looking more like a child's bedroom full of toys rather than a natural biotope! You should try to keep a cephalopod's brain occupied however you set up its aquarium.

There are lots of scientific papers written about enrichment for animals in captivity, and it really is worth your time to browse the Internet for them. Some are very relevant to cephalopods. Enrichment will be discussed further in Chapter 5.

TANK CONTENTS

As well as all the normal paraphernalia associated with aquaria, the items you put inside a cephalopod tank can fulfill the necessary role of providing some enrichment for your animal and can be as simple as adding sand for a burrowing species. But before we dive into the fun and games of amusing an octopus, some basics have to be covered first.

The Substrate

The substrate you decide to include in your cephalopod aquarium should be given careful consideration before being added to the tank. There should be enough to cover the bottom of the aquarium, and if a burrowing species is kept the substrate should be deep enough to allow it the opportunity to dig. The substrate should not have sharp edges, as a cephalopod's skin is fairly fragile, and it should not contain any pollutants or properties that will adversely affect water quality.

There are two main types of substrate commonly used in aquariums: gravel and sand. The distinction is based on the size of the particles, and each type has different properties.

Why Substrate Matters

- It decorates the tank in a pleasing manner, providing a more natural background for a mainly benthic (living on the sea floor) animal.
- It can be effectively used as a buffer to help maintain a stable pH.
- It is home to countless millions of beneficial bacteria that will aid filtration.
- Cuttlefishes and some octopuses will frequently bury themselves in the substrate as they do in nature.
- A substrate in a cephalopod aquarium makes the animal feel safe and secure when compared to an aquarium with a bare glass bottom.

This substrate was collected off a beach on the west coast of Scotland. It is a mixture of many different types of shells including those from mussels, clams, and oysters, and small stones. There are larger shells mixed in that give the substrate a very natural look.

Gravel

This has been the most commonly used substrate for freshwater and marine aquariums all over the world for many years. In fact, almost everyone who sets up their first tank will have had a layer of gravel. There is no such thing as a standard type of gravel, so as long as it is aquarium-safe and cephalopod-safe you can use gravel as your taste requires.

The best type of gravel to use for cephalopods has rounded smooth edges and a size of between 3 and 6 mm (about one-tenth and one-fifth of an inch). It is available in all aquarium stores. Most gravel consists of very natural-looking shapes and sizes; more gaudy dyed colors are available if that interests you. This gravel is inert and will have no effect on water parameters.

Gravel—and all other substrates for that matter—needs to washed thoroughly before being added to a cephalopod aquarium. It should be the first thing added to a new tank, even before the water. It needs to be washed because it will be full of dust and dirt and other possible contaminants. The easiest way to wash

Tainted Gravel

Coral gravel collected from a beach may contain metal fragments. These could potentially poison and kill a cephalopod (or other reef invertebrates for that matter). There is an easy technique for removing the metal, and all it requires is a plastic bag and a large magnet, such as the kind used for cleaning aquarium glass. Simply put the magnet in the bag and move it through the gravel. Assuming that whatever metals are in the gravel are ferrous, they'll stick to the magnet through the bag. Just wrap the bag around the fragments, remove the magnet, and you're good to go!

aquarium gravel is to use a clean bucket that has never had chemicals used in it. Add a few handfuls of gravel. Run this under a tap or hose, stirring all the time with your hand or a wooden spoon. Rinse and add more water until the water is clear in the bucket. Wash all of your gravel in this manner.

Gravel should be added to a depth of approximately 1 inch (2.5 cm) for most cephalopods, unless a particular species requires more.

Coral gravel is another type commonly sold for marine aquariums. The gravel particles are the remains of dead corals and collected from beaches near coral reefs. The gravel should be washed and added to the same depth as mentioned for regular gravel. Coral gravel will buffer your aquarium's water. This will help prevent a pH drop over time.

In the case of burrowing species, coral gravel should be avoided, as it can have sharp edges. In this case a mix of fine sand and rounded aragonite should be used.

Sand

A natural-looking bottom covering for the aquarium can be created by using sand rather than gravel.

A type of sand called aragonite is popular with aquarists because it helps keep the pH steady in the aquarium. Aragonite is made up of calcium carbonate ($CaCO_3$) and is the major component of mollusk shells.

Aragonite sand is a great choice for your cephalopod's substrate.

The Advantages of Live Rock

- Live rock can be arranged to provide the nooks and crannies cephalopods use as hiding spots.
- It comes full of beneficial bacteria and provides biofiltration.
- The rock will introduce all sorts of reef invertebrates such as worms, copepods, and amphipods, which help clean up uneaten food.
- It makes the tank look much more realistic and interesting.
- Its complex, irregular surface provides an excellent surface area for colonization by bacteria and provides a home for small reef invertebrates.
- It's fun to see what develops on the live rock. This can occur over a long period of time—new things will continue to appear even a year after the live rock is set up.

Although deep sandbeds are popular with many marine aquarists, the digging most cephalopods do makes it impossible for a biological sand bed to work properly. It is better to provide a shallower substrate so it is easier to keep clean.

Live rock constructions will provide a den for an octopus. Live rocks will also serve as a medium for beneficial bacteria.

This photograph was taken using a red light and captured a tiny hitchhiker octopus, *Octopus mercatoris.*

One way in which you could use a deep sandbed effectively is if you employ a refugium. A refugium is very similar to a sump in that it is a body of water external to the main tank but connected via an overflow system. The refugium, as the name suggests, provides a refuge for micro invertebrates, algae, and a sandbed without the disturbance of a cephalopod digging through it.

Live Rock

This is something of a confusing name for a popular item of aquascaping. The name refers to the fact that the rock is sold with various species of marine life growing in and on it. Live rock is harvested from tropical oceans potentially thousands of miles from your aquarium. It consists of pieces of dead stony coral colonies that have naturally become encrusted with algae, coralline algae, and bacteria that colonize the porous materials. Invertebrate hitchhikers are very common. In fact, many octopus keepers' first experience occurs when they add live rock to a tank and a tiny octopus appears! It happens quite regularly.

Other Rocks

The only criteria for using other types of rock are that they release no toxins and have no other adverse effects on the water parameters. Stick to rocks sold for use in marine aquaria such as tufa or ocean rock.

Flower Pots

Clay flower pots that are new and have never been in contact with any pesticides or fertilizers are great for cephalopod tanks. Many octopuses can squeeze through the drainage holes in an upturned pot. For larger individuals, cut out a hole and sand it down smooth to make an entrance hole that won't damage an octopus's soft skin. Old-fashioned clay drainage pipes are also ideal, or look for the specially made clay pipes being sold for breeding plecos—freshwater suckermouth catfishes.

If you decide to use a flower pot in your cephalopod's tank, be certain it is brand new. Gardening chemicals can be deadly.

PVC Pipes and Fittings

Octopuses have an amazing ability to squeeze through tiny gaps, and in order to feel safe and secure in an aquarium they need a place in which to take refuge—the tighter the space the better. Many octopuses will choose a plastic pipe for their den and will find suitable small rocks and shells to fashion a door.

Just like any other item added to the tank, the pipes must be safe for the purpose. You need to use the ones listed as being of food-grade quality (safe for the transportation of human food). This is because cheaper plastics can leach chemicals, which may kill cephalopods. Again, check with your aquarium store for safe pipework and use your imagination. Feel free to experiment with elbows, joints, and even sections of clear pipe.

Shells and Barnacles

The addition of mollusk and barnacle shells can help make a tank more natural looking. They also provide an octopus with hiding places and tools. Many octopuses will use shells as doors for their dens, and small octopuses will often choose a shell as a den. You could collect your own shells for your tank—scrub them clean and boil them for a few minutes to make sure they are safe.

Toys

Many aquarists would frown at the idea of adding such things as plastic building blocks or balls to an aquarium, and the thought of making it look any less than a natural biotope would make them balk, but there is nothing that says an octopus tank can't be fun!

Most aquarium stores will stock a range of artificial decorations for a marine tank. There are some quite realistic replicas of live rock starting to appear on the market as well as the good old-fashioned fun items like shipwrecks, skulls, treasure chests, etc. An octopus won't give a second thought to making its home inside a gaudily colored sunken galley, so it's down to your personal tastes there.

Plastic toys for babies and young children are designed to be nontoxic and safe in case of ingestion. It is very unlikely that these toys will release poisonous chemicals. Toys that have been used with success include small and large plastic building blocks, baby teething rings, toy balls for babies, flexible plastic

Keep Your Equipment and Your Cephalopod Separate

The art of aquascaping an aquarium provides, under normal circumstances, the opportunity for hiding essential tank equipment. Heaters and filters are normally hidden behind strategically placed rocks or plants and kept out of sight. In an aquarium meant for octopuses you should ideally have all your equipment in a sump tank, because an octopus will take great delight in taking filters apart. From a safety point of view, the heater should be in the sump, otherwise it might burn the octopus. Not having all the equipment in the tank actually makes the aquascaping of a cephalopod aquarium that much easier.

animals sold as fishing lures (hooks and all metal removed), and jars that can contain crabs for a treat. (See Chapter 5.)

HOW TO ASSEMBLE YOUR AQUASCAPE

The last thing you will ever want to hear when you are the proud owner of a cephalopod aquarium is the noise made by a rock slipping and smashing through the tank's glass. This may sound a bit extreme, but in fact it could easily happen if you don't pay particular attention to how you aquascape your cephalopod tank, especially if it's for an octopus.

Octopuses could be described as frustrated interior designers. They will arrange an aquarium to suit themselves. This means building their own caves and generally wrecking all the work you put into the aquascaping. Although this is great for the octopus and is a great enrichment, these renovations need to be controlled to prevent an accident.

If you have a large species of octopus, you might consider gluing the rocks together. Ideally, the rocks should be cemented together using a compound sold for gluing coral fragments to rocks; this glue is inert and totally safe to use in aquariums. These compounds are able to set under water, but really the gluing is best done outside the tank and with a bit of forethought. Before you add all the rock to your tank, get an idea of how the bits will fit together and add them to the tank in that order. Do this for all the larger bits of rock, or those that are to be set higher in the tank. Leave small rocks, shells, and bits of pipe unattached so the octopus can move them.

A cephalopod tank will essentially look like a cross between a fish only with live rock (FOWLR) tank and a reef tank. This is because there will normally be only one mobile animal, live rock, and few sessile invertebrates within the tank.

Adding shells to your aquarium will benefit your cephalopod.

MATURING THE TANK

It is vital that a cephalopod tank be fully cycled and mature. A cephalopod will produce approximately three times more ammonia than a fish of a similar mass. This is a critical consideration for such issues as filter size, protein skimmer size, and the amount of live rock to be used. A cephalopod aquarium should be 100-percent ready for the animal before you add it.

Set up the aquarium with water, substrate, rocks, and all other aquascaping materials, and turn on all the equipment. Monitor this for a day or two to ensure that the equipment works properly. The protein skimmer will probably not work initially and will often require a bit of tinkering to get going.

Use quality test kits to achieve a baseline level and starting point for ammonia, nitrite, nitrate, pH, and salinity levels.

If using freshly imported or immature live rock, let the aquarium go through the normal maturation process. This may take a few weeks, and if you haven't cycled a saltwater tank before, you'd be well advised to refer to some books on establishing reef aquariums.

Inexpensive damselfishes are commonly used to cycle marine tanks. However, they are notoriously territorial and may attack your cephalopod when you add it once the tank is fully cycled. Pictured is *Chromis cyanea*.

This *Octopus bimaculoides* (affectionately known as Mr. Octopus) is crawling out of his tank (left) onto his owner's arm for a controlled excursion. You should not let your octopus fall (right), but if it does happen, often the octopus is uninjured.

The maturation process normally involves the appearance of one or two species of undesirable algal outbreaks. These will normally pass with time; although they're unsightly, they are just using up excess nutrients.

Once your test kits show zero for both ammonia and nitrite, cleanup animals such as algae-eating snails and hermit crabs can be added.

In order to have an aquarium's biological filter ready for a cephalopod, the tank's beneficial bacteria and its associated equipment have to be at a level that will not produce a large poisonous ammonia spike when the octopus is added. One of the easiest ways to do this is to start stocking the tank with fish. If you intend to start with a very small species of cephalopod or a baby, only one or two fish are enough to cycle the tank for a similarly sized animal. A larger octopus will require the tank to handle a larger bioload and therefore should have several fish living within it for a period of time to build up the capacity.

The fish should be of the hardy species often sold for new tank setups. Damsels are ideal for the purpose of increasing the bioload of the tank and should be added carefully and slowly following normal acclimation methods. They should be added over the course of a few weeks and in such a way as to cause them no harm. Fishless cycling is becoming more popular with fish-only setups and with reef tanks and may work for a cephalopod tank with proper provisions; since you want a mature setup, you can introduce fish after the tank has cycled so that the biofilter will remain functional until it is time to introduce the cephalopod.

On the days building up to the octopus's arrival the tank should be prepared. This includes making the tank escape proof. (More on this in a moment.)

The fish should be removed as close to the cephalopod's arrival time as possible. They have served their purpose here and can either be moved to a separate tank or possibly back to the original store they came from. Catching the fish may be a horrible task because of the aquascaping, but it is absolutely essential. This is because we are trying to avoid any spikes in ammonia that would occur if the fish were not removed first, since the cephalopod's wastes will be added to theirs, and the biofilter will not be able to handle the load.

Do not feel tempted to leave the fish with the octopus or any other cephalopod. In some cases the cephalopod will hunt down and eat the initial fish, but it can equally go horribly wrong and an octopus can be attacked, hurt, and even possibly killed by an existing fish, especially territorial fish like damsels. Err on the side of caution and remove the fish.

Test water parameters daily and pay close attention to ammonia, nitrite, and pH for fluctuations. A slow and steady maturation process will prevent problems.

The octopus might disappear into the rockwork for a few days. If it is a nocturnal species you may need to wait a long time to catch a glimpse, but diurnal species will explore during daylight hours and eventually make an appearance. Do not start removing rockwork and digging about to find your new aquatic pet. Do not strip down a tank or make a tank bare so you can see the animal. You will hinder its adjustment to

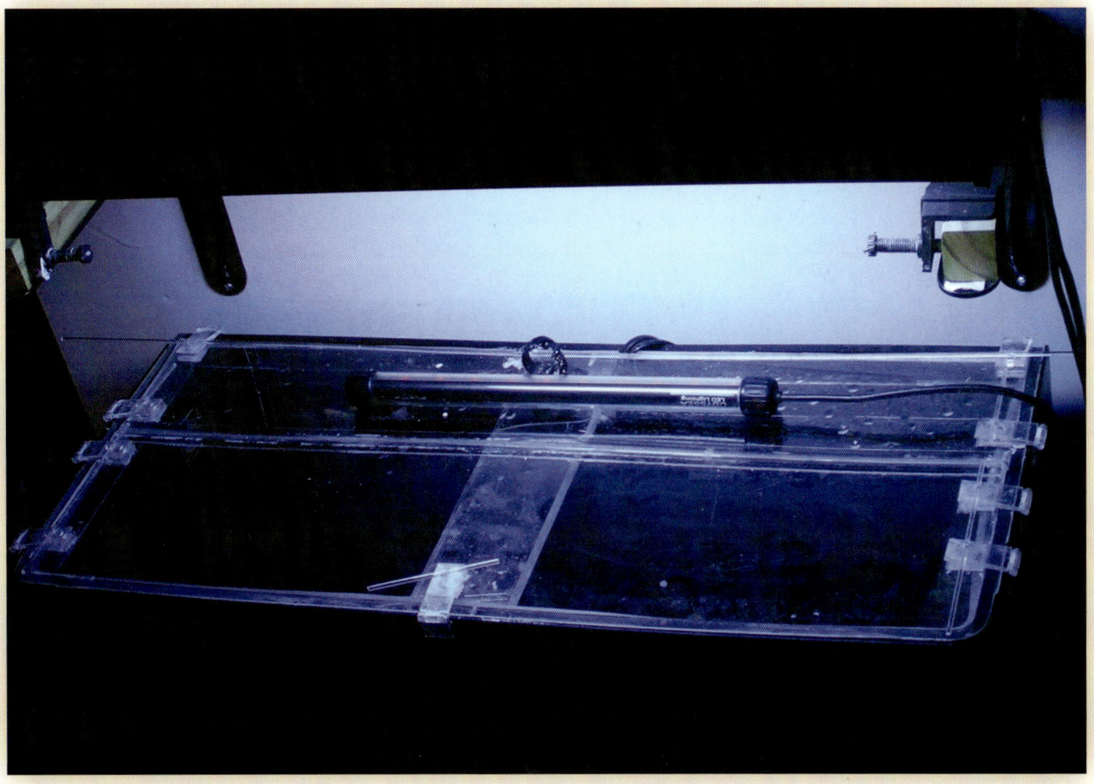

A secured cephalopod tank. The plastic clips prevent the octopus from being able to escape.

aquarium life by doing so and probably decrease its lifespan, especially if the cephalopod is a baby. Young cephalopods naturally go through a shy hiding stage to avoid detection by predators.

HOW TO MAKE YOUR AQUARIUM ESCAPE PROOF

This topic is most relevant in regard to octopuses, as these animals have perfectly evolved into powerful eight-armed, sucker-laden, elasticized escapologists. They can climb glass, squeeze through tiny holes, leave tanks via overflows, and if all else fails, open the hood and simply climb out. Worse yet, they are illusionists who will leave you with no doubt that they have climbed out of the tank only to appear moments later, still in the tank sliding out of an impossibly small cave. How many times have both authors stared into an empty-looking aquarium and not seen the octopus only to realize that it was perfectly camouflaged and staring right back at them the whole time? Let's just say more than once.

If you are using a system with a sump there are two main escape routes—the intakes and outlets that take water to and from the aquarium. Thankfully these can quite easily be safeguarded against escape by covering them with either a plastic grill or a sponge made especially for the task. Many filters now come with such attachments. DIY options are available; for example, you can use plastic mesh siliconed in place at overflows.

The hood (top) of an aquarium also offers some escape routes—you should pay attention to holes normally utilized for power cables and pipes. They should be sealed with an aquarium-safe silicone sealant and left to cure for the recommended period of time. Duct tape can also be used but must be replaced from time to time.

The way in which the hood sits on top of the aquarium requires some attention, as hoods do not always have a tight fit and in some cases can be easily moved. It is impossible to cover every make of aquarium hood in this book—when you are in the planning stage before buying your aquarium, think ahead to sealing up unnecessary holes. Duct tape is your friend! It may not be particularly attractive but if all else fails, you can tape down hoods and lids. Just remember not to make it permanent.

Do your best to make the aquarium as escape proof as possible and don't panic if you can't see your new octopus. A well set up and aquascaped tank should allow an octopus to take refuge and hide whenever it wants, especially in the first few days.

Cuttlefish Aquascaping

The aquarium layout for cuttlefishes can be very much like a typical reef tank decorated with a layer of fine sand and live rock for refuge. Remember, you can keep cuttlefishes singly in tanks or as groups in very large tanks of a few hundred gallons in size. *Sepia bandensis* does well in tanks with lots of live rock. If you're keeping a group of this species, always ensure that they have many suitable hiding places and possible territories. In the few recorded attempts of keeping adults together permanently, the results were fighting and deaths, but this could be because of a lack of space. With multiple *S. officinalis*, you can have a rocky mound forming some caves in the aquarium, but since this species gets so large you might want to consider removing some of the rockwork and just leaving them with one cave each to free up some swimming space. Large clay plant pots are a good substitute for rock caves.

Cuttlefishes are not the escape artists that octopuses are known to be.

Escape Proofing the Cuttlefish Aquarium

If you are planning to keep a cuttlefish you do not have to worry quite as much as you would with an octopus. A cuttlefish will not climb out of a tank like an octopus, but cuttles can potentially propel themselves out of the water if frightened. A hood should still be used with a cuttlefish, although it doesn't need to be as tight-fitting.

A cuttlefish grabbing for food with its arms will sometimes catch hold of the aquarium glass near the water's surface and then use its siphon to expel water as if to shoot off backwards. What actually happens is that it ends up spraying water—sometime a *lot* of water—out over the side of the tank. One of the authors of this book, Colin, had two experiences with such a situation, both involving the same adult male *Sepia officinalis*. On one occasion the cuttlefish grabbed a cross bar used to support the aquarium's hood and sprayed a jet of water over the author while he was carrying out tank maintenance—holding an electric device. Luckily there was no serious shock! On another occasion the male cuttlefish sprayed a jet of water and ink all the way down the living-room wall. Again the author narrowly survived with his life—this time from his spouse!

There are many stories about long-term captive cuttlefishes spraying jets of water at their caretakers, especially in large open tanks where they are kept for breeding. Use a hood to prevent them from spraying you with water and ink.

Chapter 5

Introducing the Octopus or Cuttlefish into the Tank

So the big day has arrived! You ordered your cephalopod and it has either arrived at your front door all safe and snug in a box or you have just come home from the aquarium store with one in a bag or bucket. Now what?

ACCLIMATING A CEPHALOPOD

A day or two before the cephalopod's expected arrival, you should mix up some new aquarium water and keep it well aerated in a bucket and at a similar temperature to the water in your aquarium. You are going to need this to replace water used during the acclimation process.

Ideally you should set aside a few hours for acclimating your new pet—the process doesn't stop as soon as the cephalopod is in the water!

No matter whether the cephalopod has arrived in a bag within a box or in your own bucket, there are several key parameters that must be equalized gently to cause the least amount of stress on the animal and to prevent an inking episode. The main issues are light, temperature, and the general water quality.

Light

Your octopus is probably wrapped up well to conserve heat and, because of the container's insulation, it is in the dark. It could have been this way for more than a day if it was sent to you via courier. If you suddenly open the box and flood the interior with light, it will scare the octopus and possibly make it ink in the bag. Although the ink isn't toxic itself, in a small amount of water it can quickly coat the gills

Just arrived: Four carefully packed cephalopods.

Placing a bagged octopus into your main tank will adjust the bag water's temperature, but it will not acclimate your cephalopod to your tank's chemistry. You must acclimate your new pet properly or else the chemical shock can harm or even kill it.

and lead to breathing problems. Inking should be avoided at all costs during the acclimation process. Remember, it is a sign of stress, and you want to minimize stress on the animal.

You should carefully open the lid of the box, remove the polystyrene lid (if there is one), and expose the interior bag just enough to let some light in. After five minutes or so, you can open the box a bit more. Once ten minutes or so has passed, you should be able to open the box entirely. (Most buckets used are made of white plastic, so a decent amount of ambient light should still get into the bucket from the room.) It is a good idea to work in a dim environment with the tank's lights off. Leave the bag tied at this point (if applicable); a bucket should not be left uncovered, as there is the risk that the octopus will climb out.

In the event that you open the box and find that the octopus has inked in the bag and you now have a large polythene sack of jet black water in your hands, you can speed things up a bit. This is for emergencies only, but if you can't see any signs of life in the bag you should act even more quickly and get that bag open. If you can see movement, whether it is an arm twitching or the whole octopus moving, you

A newly arrived *Abdopus aculeatus* ready for acclimation.

should follow our recommended acclimation process and resist the urge to speed things up too much. By all means open the bag and add some tank water, but do not add too much.

Stay!

Some octopuses, especially larger species, will show their discontent with the acclimation process by immediately trying to climb out of the bucket. This behavior must be prevented. You should use a bucket with a lid; the only opening in the lid should be just small enough to fit the piece of airline tubing.

Water Quality and Temperature

At this point you can decide whether to acclimate the cephalopod in the bag or whether to transfer the entire contents of the bag into a bucket with tight-fitting lid. Assuming that the octopus and its water have been poured into a bucket, it can make things easier if you add something like a small plastic cave or pipe—the octopus will normally climb inside during acclimation and will ultimately be easier to relocate by lifting the cave with the octopus still inside into its new aquarium.

You should remove a cupful of water from the bag/bucket to test the water's parameters. Test for temperature, pH, ammonia, and salinity. Ammonia is less toxic at a low pH, so you should try to avoid a sudden pH change at all costs. These test results will give you an idea of what the octopus has been sitting in, and they will highlight just how slowly you should acclimate it when compared to your aquarium's water parameters.

No Bucket? Bag It!

If you don't have a bucket you can acclimate the cephalopod in its transport bag using the drip method, but use elastic bands to reseal the bag after each opening to ensure the cephalopod stays put. A clothespin can be used to attach the bag to the side of the aquarium, hold it closed, and stop it from drifting away if you want to put the bag in the aquarium during acclimation. You can do this for very small octopuses and add water with a clean glass. Just be sure not to dump all the water from the bag into the tank at the end.

The best method of acclimation to use is called the drip method, adding water from your aquarium to the bucket/bag one droplet at a time over a long period of time to avoid any shocks in conditions. Start a siphon from the main tank through a length of airline tubing. Control the flow with either a clamp or a valve (or simply tie a knot) so approximately one or two drips per second are released. This ensures that the octopus has plenty of time to adjust to any changes in temperature, salinity, and pH.

Once the temperature, pH, and salinity are all close, if not the same as the main tank, and after at least an 80-percent addition of new water from the main tank, it is time to transfer the octopus into the tank. Be sure to add as little of the transport water as possible to the aquarium; even better, avoid adding any at all. This water is less than perfect as a result of nitrogenous waste and will deteriorate your aquarium's water chemistry.

Hopefully the octopus has decided to hide out in the "cave" you provided and you can transfer it into a jug and lower it gently into the tank. Try to do this while keeping the octopus underwater at all times. Keep your hands clear of the octopus while doing so.

If the octopus is not in the hiding place you can slowly and carefully remove water from the bucket until the octopus is just covered. Next, tip the bucket and its contents into your tank. The octopus should slide right in, but some are more stubborn that others!

Use the water you mixed up prior to the octopus's arrival to top off the tank if necessary, and seal the cover. Over the course of the remainder of the day leave the lights off and do not disturb the tank. Do *not* feed the animal!

A newly arrived cephalopod is going to be scared and unsettled. The last thing it needs is any more disturbances, so don't introduce the new pet to friends and family just yet, no matter how exciting it all

Acclimating Cuttlefish Eggs

If you received cuttlefish eggs, you should follow the same slow acclimation process. Most eggs are best hatched out in a net breeding trap like the ones sold for freshwater livebearing fishes and kept there for at least their first few feeds. The eggs will hatch at different rates and over the course of several days.

seems (besides, more often than not it will disappear into the rockwork anyway). What the animal needs is peace and quiet. Time is the best healer for an octopus that has just arrived. Start the tank regimen with the normal photoperiod the next day and continue from there.

Cuttlefishes can be acclimated in exactly the same way, but there's no risk of their climbing out of the bucket. Transfer cuttlefishes in a jug and avoid exposing them to air—remember, this makes them ink. If a cuttlefish is being transported in a container with hard sides it is important to avoid sudden knocks and bumps, as cuttlefishes tend to jet off backwards when frightened and damage the back end of their mantles. This is commonly called "butt burn," but rather than a burn it is damage to the mantle and its inner shell. In extreme cases the cuttlebone can be exposed.

The transport bag may have had old disc covers from the cephalopod's suckers floating in the water. This is nothing to be alarmed about; cuttlefishes shed discs on a regular basis.

THE FIRST FEEDING

As tempting as it may be, do not feed a newly introduced cephalopod on the same day you get it. Many will not eat, and uneaten food in the tank will rot and foul the water. It is best to leave the animal unfed its first night and try the next day. Don't worry about starving it; an octopus can live many days if not weeks without food.

Cuttlefishes are quicker to settle in to a new aquarium—they are generally diurnal and hunt through the day, relying on sight. Like octopuses, however, they will usually show interest in live foods only, as live foods make up their diet in the wild. Dead items just do not look like food to a cephalopod and should be introduced only after the animal is settled and feeding properly. This is why it is so important to have a source of live food ready *before* you acquire the cephalopod.

An octopus is likely to be a bit shyer than a cuttlefish. Many are nocturnal, and even diurnal species will be less likely to forage during the day under the new confined conditions of your tank. In that case you can try to feed it for the first time with the tank lights off, with only ambient light from the room. Here too live foods should be used, especially crabs. Drop the crab in the octopus's line of sight then sit back and observe. Your forward-facing eyes make you look like a big predator to an octopus. It will be some time before your octopus learns to trust you. This will come in time but can't be rushed.

Numerous Nifty Names

Of all the cephalopods, octopuses are the most likely to be given individual names and to exhibit individual personality traits. Even in public aquariums, where many of the animals go unnamed, octopuses often have names. What names have been given to octopuses kept in home aquariums? There's quite a variety. Some have names related to being an octopus, like Octavia, Pulpo, Ink, and Prowler. Names starting with the letter "O" are popular too: Ollie, Otto, and Octavia. Others have classical names, such as Polyphemus, Medusa, Gaius, and Achilles. Trapper was named because she was caught in a crab trap. Thomas, Lindsey, Herman, and Einstein have human names. Then there are Biddle and Fluffy and the more fanciful Glove and Octane.

It can happen that the sex of an octopus becomes apparent only after eggs are laid. Just like that Conan became Conanny.

Your new cephalopod will be shy and will hide for a day or so. Don't be alarmed if you don't see it. Eventually, though, it will come out and explore its new home.

Foods to Use

As already mentioned, the best food for newly introduced cephalopods is live food. Most cuttlefishes will ignore dead food, and many octopuses will too. Newly hatched cuttlefishes will spend a couple of days using up the remainder of their yolk sacs before being ready to feed. It is absolutely imperative that they are offered live foods—amphipods are a good choice—or they will starve to death.

The best choices for any newly introduced cephalopod are either crabs or shrimps, as those animals make up the bulk of a cephalopod's diet in the wild. Hopefully, your established aquarium will have a good source of home-bred amphipods. Octopuses will always hunt them down.

With luck you will see the cephalopod eating, but it is not a disaster if it doesn't for the first couple of

Have You Seen My Octopus?

A day-old *Octopus briareus* was added to the tank of one of the authors of this book (Nancy) and disappeared for four months before being seen again. Of course, it was assumed to have died. One day it just suddenly appeared. It had managed to stay totally hidden in a 19-gallon (80-liter) tank in plain view. The small *O. briareus* had lived entirely on the food available in the tank—amphipods, snails, and crabs—with no additional food items added. In that time its mantle size had increased from a quarter of an inch to 1½ inches (from about 6 mm to 4 cm).

days. Peace and quiet and a choice of live foods will get it to eat. The fewer disturbances caused by you the better, so hands out of the tank, please.

BEHAVIOR IN THE NEW TANK

The behavior shown by a cephalopod in a new tank will vary from species to species and indeed between individuals. Some species, such as *Abdopus aculeatus,* will often be found hanging from the front pane of aquarium glass from the first day. Others, such as *Octopus bocki,* will take up residence in a barnacle cluster or coral head and vanish for days. Either way it will normally take days to weeks before the octopus or cuttlefish adapts to aquarium life and begins showing normal behavior.

THE "BABY HIDING" STAGE

The "baby hiding" stage is a term coined for those hobbyists lucky enough to get a very young octopus. A baby octopus in the wild is potentially a tasty morsel for almost any other animal in the sea, including other octopuses! They must use every adaptation they have to survive in such a hostile environment. The easiest way for a baby octopus to stay safe is to stay hidden: it either spends a lot of time in small rocks and caves feeding under the cover of darkness or uses its excellent camouflage to blend into its surroundings.

If you have a baby octopus, do not fret if you have not seen it in the tank for a few days, or even a couple

Offer your octopus plastic building blocks and shells to play with. In this photo, young bimacs are peeking out from their barnacle homes.

of weeks. Chances are that it is still there despite your unsuccessful scanning the tank for signs of life. There might be a beady eye or two on you at all times—you just haven't spotted them yet.

Young octopuses will feed almost exclusively on amphipods in an aquarium. Your tank should already be well established with live rock if you plan on getting an octopus, especially a baby, so a good amphipod population should be flourishing. You can often find suppliers of live amphipods on the Internet to boost your population too. Also, with a little research and practice you can culture amphipods at home.

This disappear-reappear behavior leads us to a very important issue. You should never tear your tank apart if you don't see your octopus for a few days. Take it as a compliment that you have set up your tank so well that the octopus can behave like an octopus and do what it does best, which is to stay hidden and out of danger. It can be an anxious time, of course, but acclimating your first cephalopod and trying to ensure it feeds only to remove all of the rocks to check whether it is still in your tank is detrimental to the settling-in period. There have been many instances reported where people have panicked and ripped the tank apart only to find that the octopus is alive and well but now very scared. Patience is the key here, and it is ultimately worth it in the long run.

GETTING TO KNOW YOUR OCTOPUS

After the initial period of your octopus's settling in and getting accustomed to its surroundings by establishing an eating pattern and so forth, you can begin to see more natural behavior. To an extent you can start to interact with the octopus, but don't rush things too much. Allow your octopus a few weeks to feel comfortable and safe, and then to get used to your being around and providing food.

Interacting With Your Cephalopod

Cephalopods are by their very nature curious animals, and they will sooner or later become very curious about you. "You" being that strange animal that appears every so often, doesn't try to eat them, and actually drops food right in front of them! As far as the octopus is concerned, "This needs to be explored further…"

The initial breakthrough with a cephalopod is when it starts associating you with food. This is the single largest step that will be taken and signifies the start of what can be best described as true interaction between you and the octopus or cuttlefish. This is when it will stop hiding when it sees you and in fact will move to the front of the aquarium in anticipation of food when you are nearby. It will begin to show you various colors, skin textures, and even reach out for your finger if you hold one against the glass. Cuttlefishes will hold their arms in bizarre manners and have even been known to copy hand gestures to an extent. Some specimens will start to climb up the glass to the area through which you drop in food and will even squirt jets of water out of the top of the tank with their siphon. All in all, this is when things get interesting.

Can I Touch My Pet?

Bearing in mind that octopuses and cuttlefishes may try to bite you, you have to exercise caution when touching them, and doing so certainly isn't an option for the more venomous species like blue-ringed octopuses in the genus *Hapalochlaena*, which should never be kept in a home aquarium. Most bites are associated with exploring your hands, but even common species like *O. briareus* and *O. bimaculoides* may

An octopus will adopt a corner or crevice as its den and will spend a lot of time in it.

leave your hand with a stinging sensation and may even break the skin. You certainly wouldn't want a bite from a large octopus like *O. vulgaris* or *O. cyanea,* as the sheer size of their beak means a larger and more powerful—and potentially more harmful—bite. Any octopus can envenomate you through its bite, though not to the rapidly fatal extent of the blue rings.

However, if you touch an octopus's arm or part of its mantle during feeding or playtime this will not hurt the octopus as long as your hands are clean, free from chemicals, and you lightly touch the octopus without holding it. Some octopuses seem to enjoy the interaction and even like being rubbed lightly between and above the eyes. Cuttlefish shouldn't be touched, as they can become frightened and jet backwards and hurt themselves.

Can I Take My Pet Out of the Water?

This is another question that is asked by new cephalopod keepers from time to time. Many think of this because an octopus used to being fed will often try to climb out of the water to get food from you rather than wait for it to be dropped in. It will poke arms out of the water and even try to climb out onto the tank braces and hood. This is a dangerous position for an octopus to be in—a fall might injure it.

Octopuses like these can be fed above the water line if you like, but their safety comes first. Your own safety should be considered, too, as there is likely to be some splashing and a lot of electrical equipment nearby.

Dens and Hidey Holes

Cuttlefishes will not be so bothered about having a den or a hidey hole as long as they have somewhere they can lie down and sleep during the day, such as a sheltered part of the tank that is a bit darker. Octopuses, on the other hand, must have access to a variety caves, rocks, and pipes in which they can establish a den.

Most wild octopuses will utilize a single den or a series of dens within their territory in which they can seek refuge from a predator or to rest. The size of the den is important; they would normally squeeze into as tight a fit as possible to feel secure. You should always have a selection of dens available in your aquarium, especially as your octopus grows. Some misinformed hobbyists make the unfortunate mistake of having a sparse tank in the hope that they will see the octopus more frequently. In reality, though, the opposite happens. The octopus will feel less secure and is less likely to behave in a relaxed manner; it will hide all the time. The bigger the selection of caves, dens, and pipes available the more relaxed the octopus will feel and ultimately the more you should see it.

Stimulation and Enrichment

After water quality and feeding, the stimulation and enrichment of a cephalopod's life is most important to success. Enrichment for a cephalopod can take many forms, and you should try a combination of these forms for your pet octopus. Cuttlefishes will show less interest in manipulating toys, but most of the points that follow will apply to them as well.

What is Enrichment?

As we mentioned earlier, enrichment is a way of enhancing a captive animal's environment by providing means of stimulation. In other words, an octopus kept in a tank might become bored because it no longer has to hunt for food, watch out for predators, or do what it normally does in the ocean. Research has shown that octopuses spend a large percentage of their time in their dens, safe from predators. But the time they spend foraging and hunting enables them to use their intelligence and gather information in ways that may not be replicated in the aquarium.

What are ways of providing enrichment for such an intelligent animal? Some researchers feel that providing an environment close to the one the cephalopod came from and putting live food in the tank may be enough. When live crabs are offered, the octopus must use its skills to hunt and capture them,

On The Hunt

Sometimes an octopus has to learn to hunt a particular species of live food. One of the authors of this book, Nancy, had a wild-caught three-month-old *Octopus bimaculoides* who knew how to hunt crabs, snails, and even fishes, but small shore shrimps from Florida were a complete mystery to her. Ollie's first attempts at catching these 1-inch (2.5-cm) crustaceans involved jetting after them head first, then trying to reverse and catch them in her arms. Needless to say, the shrimps were far too fast for this method of capture. She tried various methods and ended up hunting them (which she loved to eat) at night when the lights were off, having turned completely black, and hanging upside down at the water's surface. This technique worked perfectly because the next morning the tank was shrimpless. This shows how she tried various strategies until she found one that worked, and how the little shrimps engaged her and caused her to react to her environment.

Plastic toys sold for human babies can be offered to your cephalopod.

even more so if a few crabs are released and have time to hide. Live shrimps and even crayfishes offer a challenge, too.

It's also possible that having other marine inhabitants to watch and having some other species in the tank (a brittle star, some small shore shrimps and crabs, an urchin) will also contribute towards enrichment. Whether any enrichment is provided by watching people move about the house, watching mobile pets such as a cat, or even watching sports and cartoons on television, all things an octopus may do in the home aquarium, has not been researched, though.

Octopus Toys

The most attention-getting method of enrichment challenges octopuses through toys. They're good at opening puzzle boxes, finding food hidden within plastic containers, and even taking apart hollow plastic toys to get to a treat inside. You just need to find the appropriately sized toy that suits your octopus.

Experience shows that different octopuses (individuals, not different species) react differently to toys. Not all octopuses are intrigued by toys, and some prefer to play games with their owners. Others, especially certain species of dwarf octopus, are completely uninterested in toys or games.

When choosing a toy, make sure that it's plastic and has no metal parts or hard jagged edges. Toys suitable for human babies are good, as are teething rings. The little plastic round containers that toys come in from vending

One Toy to Avoid

Avoid ping pong balls, as they could harm your octopus when they break.

machines work well too. Many octopuses are quite taken by plastic building blocks, plastic balls, and similar toys (an octopus well known to the authors of this book, a bimac named Ink, even floated around the aquarium on a large building block). Another idea is to use soft plastic fishing lures, which come in many marine shapes. All hooks and metal must of course be removed to make them safe for octopus play. So far, these soft plastic shapes have been favorites for some octopuses. Yet another idea is to construct a little toy out of drinking straws, a simple toy that has been tested and accepted with enthusiasm by octopuses.

It's all the better when you participate in the play. A feeding stick can be a wonderful toy and good for tug of war. Sometimes, though, you may find yourself as an unwilling participant in your cephalopod's fun. Many octopuses discover the algae-cleaning magnet if it's kept in the tank. Sooner or later, your octopus will pull away the part of the magnet inside the tank, causing the outer part to fall. So the octopus has captured a good toy to play with, and the loud bang that the outer part makes when it hits the floor usually attracts you. They learn quickly that you can be summoned by pulling the cleaning magnet free! Some playful octopuses are reluctant to give up the magnet, even to the point of dangling it temptingly in the open water and jerking it back into the den when you try to retrieve it.

These are just some ideas for toys for your octopus. Many others can also be used; just make sure you avoid anything metal or any plastic that could break (octopuses are very strong, even when they're young), and experiment to see which toys are the best received. And remember, if it's sold for human babies, it can be good for your octopus.

This *Octopus bimaculoides* has learned how to open a box to get at the delicious crab inside. With practice you can teach your octopus how to do the same.

Opening Jars

Now that a number of public aquariums have octopuses that open jars to find a food treat, people who have pet octopuses wonder whether theirs can do this as well. They can, but you first have to teach them how.

First, start with a jar of an appropriate size for your octopus. There are many videos posted on the Internet of small octopuses struggling to open large jars—do not do this, and use a jar that the octopus can wrap its arm-webbing around. The jar and lid should both be plastic, both for safety compared to glass and so you can drill holes in the lid. Drilling holes allows the food item's scent to enter the tank's water to entice the octopus. Place a crab or a shrimp in the jar, fill it with tank water, and lay the lid on top of the jar, turning it just enough so that it is still loose but not likely to float off when you add it to the tank. Each time your octopus figures out how to remove a loosely fitting lid, screw it on just a tiny bit tighter and keep screwing it on more tightly as your octopus learns to handle the process. Soon you will have an octopus that opens jars. This whole process, though, often takes many days, weeks, or even months to accomplish.

Cuttlefish Enrichment

Cuttlefishes can be more reluctant to take nonliving food than octopuses, so they may be getting more enrichment from hunting their food. Since they can live in groups, they have the stimulation of other cuttlefishes. So far they have not been interested in toys but will still see you approach from a distance and will usually swim to the front of the glass to get your attention in the hope that you feed them!

Do Octopuses Watch Television?

We could not find any research done on this question, but we have received so many reports of octopuses watching television with their owners that we have to consider it a possibility. This would be a form of enrichment, much like watching fishes go by while sitting in their dens. They apparently like brightly colored programs such as sports and cartoons. So you might consider situating your tank where your cephalopod can watch television with you!

Chapter 6

Feeding Cephalopods

The information in this chapter is slanted towards octopuses since they are more commonly kept, but where necessary we have included specific information about cuttlefishes. A lot of the information is transferable between the two groups.

THINGS TO CONSIDER EARLY ON

Acquiring the correct food for an octopus can often be one of the hardest tasks in preparing to keep a cephalopod. It can be difficult to find an all-year-round sustainable source of food. This is especially true if you manage to buy cuttlefish eggs or have an octopus that lays eggs—you will find yourself with lots of very hungry mouths to feed. A cephalopod will usually only eat live foods, at least to begin with, and they need a continuous supply for proper nutrition and growth.

HOW MUCH AND HOW OFTEN?

The correct feeding frequency for cephalopods will vary from species to species and from individual to individual, and external variables like the water temperature will also have an effect on the cephalopod's metabolism (the warmer the water the more it eats). Your feeding regimen will require a certain amount of guesswork and careful observation on your part, especially in the early days.

An octopus can survive without food for quite a long period of time, perhaps as long as a few weeks, but such long fasts are not in the interests of good husbandry and should be avoided.

Octopuses and other cephalopods will eat on a daily basis in the wild because they have a rapid rate of weight gain—some species are able to convert 40 to 60 percent of what they eat into weight. For example,

Octopuses love crabs. Be sure that the crab's pincers won't be able to seriously harm your cephalopod.

Crab Percentages

If fed a crab, an octopus is able to eat approximately 50 percent of the weight of the crab. It discards the exoskeleton (and usually the legs and claws), so if a crab weighs 10 g (a little more than a third of an ounce) then only 5 g (a little more than one-sixth of an ounce) is actually available as food for the octopus to eat.

this means that for every 100 grams (3.5 oz.) of food eaten by an octopus it could put on 40 g (1.4 oz.) in weight. Furthermore, a young healthy octopus can increase its body weight each day by almost 10 percent if fed correctly.

Another factor to consider is the types of foods used for feeding cephalopods. The sheer volume that one can eat means that if an item of food goes uneaten and unnoticed it can cause major problems in regard to the water quality in a very short space of time as it rots. While this is true of feeding any marine predator like a moray eel, lionfish, or grouper, these fishes are generally much more forgiving as regards a deterioration of water quality. A cephalopod, however, tends to be much more sensitive to pollutants in the water. The octopus's messy high-protein diet is one reason why the aquarium needs quality protein skimming and filtration with regular water changes.

SOURCING FOOD

One of the more difficult challenges you may encounter when keeping cephalopods is in obtaining the correct types of food. Your food sources should be well researched before you buy an octopus or cuttlefish to ensure that they are reliable. Some specimens will take to dead food quickly, but some will never adapt to it at all, so make sure you have your live food supply sorted out before buying your cephalopod. Buying one first and hoping to sort out food second is a recipe for disaster.

It is critical that you locate, buy, and have on hand an appropriate selection of food before you bring home any cephalopod. If you purchase an octopus that you see in a tank at your local fish store, at least you know what size of octopus you're getting and what size of food you'll need. You will also be able to ask the store's staff what the octopus has been eating in the shop. However, if you order from an online source (which is becoming increasingly more common), you may not be able to get all of this information. As we mentioned in Chapter 2, some Internet dealers don't know themselves which species they are sending out, so you should be prepared for such a possible mix-up by having a selection of food of varying sizes and live crabs on hand. Even picky eaters almost always accept live crabs.

Ask About Diet

When buying your cephalopod, always ask what the octopus or cuttlefish has been feeding on in the store so you can offer the same foods in the beginning. Once it's home, move it onto a real cephalopod diet of live crabs and shrimps.

Sandy beaches at low tide may have suitable pools for catching shrimps. If you opt to collect your own live foods, be sure to check with your local government about restrictions and the possible need for licenses.

Octopus Foods

- shrimps (live, fresh, or thawed frozen)
- hermit crabs or shore crabs
- larger crabs
- scallops
- marine snails
- mussels (should be cracked)
- clams/cockles
- krill
- crayfishes/crawfishes (as a rare treat because they are freshwater and nutritionally inferior to marine foods)

Before your octopus arrives you should try to have amphipods, small crabs, and snails (already living in the tank), some live fiddler crabs (in a separate feeding tank), and some frozen raw shrimps. This is a good starting point.

FEEDING OCTOPUSES

If you manage to acquire a juvenile octopus you should assume that it is going through a rapid growth phase and ensure that it always has access to suitably sized food.

A new baby octopus will quite often disappear into the rockwork for at least a few days after you add it to the aquarium. It will probably take up a nocturnal existence and you might catch only fleeting glimpses of your new pet.

On Feeding Daily

Like most other opportunistic feeders, octopuses do not need to eat every day, but some aquarists prefer a daily feeding schedule because of the contact and interaction with their pet—feeding time is the best opportunity for your octopus to get to know you. They also experience a faster growth rate with daily feeding.

If the octopus is very small (a total arm span of less than 2 inches, 5 cm), it will be feeding on amphipods and any other small critters it manages to find in the aquarium, which is assuming that you have used live rock in the tank. If not, you will need to be using suitably small shrimps and crabs.

At this stage in the young octopus's life cycle, it can be very difficult to regulate how much and how frequently the octopus is feeding, but it is not too early to start trying to develop a set feeding time. The most advantageous reason for setting a feeding time is that you can train your octopus to be active when you are. Some species are naturally diurnal, but even many of the nocturnal species can change their sleeping habits to fit in with our periods of activity and will often be waiting at the front of the aquarium at feeding time!

In a tank so full of rocks and hiding places it is important that you can see the octopus eating and be able to quickly find and remove uneaten food. It's also important to see signs that your new octopus has eaten. For this reason, very young octopuses are often kept in a smaller plastic container (see Chapter 3) until they are established in their new homes and have eaten consistently and well.

How Much and How Often, Revisited

The frequency and amount of food offered will very much depend on the species of octopus, its size, the water temperature, and its level of activity. Therefore the size of the food item should be suitably proportioned to the size of the octopus—about one-fourth of the octopus's size and no more than one-third the size of the mantle length in the case of a live crab with big claws. Some hobbyists will declaw the crabs, but most octopuses have no problems with dealing with large claws. It's what they have evolved to do!

It might take the octopus as long as an hour or so to eat a single crab. Afterwards you should remove any and all leftovers: all those legs and claws rarely get eaten. Watch carefully and you can soon decide whether you need to feed more or whether to reduce the amount per feeding. If the octopus will take another crab straight away, let it have it and see whether you can discover just how much it can eat. Doing so will give you an average amount of food to start with. Feeding an animal in this way is called feeding it to satiation and is a common method of feeding opportunistic predators.

How to Feed Your Octopus

The safest and easiest way to feed your octopus is to open the lid of your aquarium and drop in a crab or shrimp near the front where you will be able to watch it fall. Hopefully this will encourage the octopus out of its hiding place. The prey item will normally head for the bottom of the aquarium and try to dig

This *Octopus bimaculoides* is resting contentedly in its midden—the pile of uneaten and discarded food parts.

into the substrate or hide within the rocks, though a hermit crab might be blissfully unaware and trundle across the substrate. Many feeder crabs won't make it very far after being dropped into the tank and will be pounced upon by the octopus as they float down the water column.

A new or young octopus might not react quickly, and something like a crab may manage to escape into the substrate. In this case you can leave it in the tank and assume that the octopus will find it later. If it is a crab it probably won't matter too much if it doesn't get eaten right away. It should just skulk about until the octopus sees it.

Shrimps will also live quite well in a cephalopod's tank if they can evade capture, but this isn't something you should allow for other live foods like burrowing mollusks. Clams and cockles, which are sometimes used as octopus food, should not be left in an octopus's tank past feeding time. Not every octopus will accept bivalves, especially large clams, which an octopus will have to drill into with its radula. Even cracked clams and mussels are often rejected.

As filter feeders, these bivalves are harder to keep alive in the aquarium; without special care they can die and go unnoticed, especially if they dig into the substrate. If your tests show poor water quality or if a cephalopod is not acting normally, these could be the first signs that something is wrong, and you should check for uneaten food. Keep track of what you are feeding the octopus and remove uneaten food when necessary. An octopus that has been in captivity for a while quickly learns how to intercept its food!

Any nonliving food items should be carefully accounted for and removed if uneaten after a short period of time. You can wrap nonliving food in cotton thread and leave it dangling into the aquarium so the cephalopod can reach it but only do so in such a way that you can pull it out easily without losing track of it if the octopus decides not to eat. A feeding stick or tongs is also a useful tool for spearing uneaten food and removing it from the tank.

The Reluctant Eater

If the octopus does not show an immediate feeding response, leave the tank alone for a while to see whether it feeds later. It can often take octopuses a while to figure out how to catch and eat a new food item.

There are also times when an octopus doesn't seem interested in what you're offering, so simply try offering something else. If you're feeding mainly thawed frozen shrimps, offer live crabs or live shore shrimps. Just like humans, octopuses can have strong food preferences that vary from individual to individual and from species to species. They can grow tired of a food (but usually not of crabs or shrimps). For example, one of the authors of this book, Nancy, had an *Octopus bimaculoides* that she fed mainly fiddler crabs and alternated with thawed frozen shrimps and fresh scallops. One day the bimac refused to take the scallops and actually never ate them again. Be prepared for such unexpected changes in food preferences and have an alternative food available.

This crab will not last for long once in the grasp of the octopus.

Saltwater Versus Freshwater Food Items

If the live food you're using is marine in origin it is usually safe to leave it in the tank until the cephalopod decides to eat it. If, however, you have used live food from a freshwater habitat, which is recommended only for very occasional use, it shouldn't be left in the aquarium for too long, as it will die quickly and cause major water quality problems.

Some freshwater crustaceans such as shrimps will survive in a marine environment for a few hours before succumbing to osmotic shock, but even they shouldn't be left in for too long if the cephalopod doesn't eat them immediately.

Feeding New Octopuses

New octopuses might be either crepuscular or nocturnal, so feed them at night just before lights out. Have a good look around the tank the next morning to see whether you can see any leftover carapaces or other body parts from the prey items to determine whether the octopus has fed or not.

Do not worry too much if you never actually get to see a new octopus eat for the first few days or even weeks. They can be very secretive while they adjust to aquarium life. Don't dig about in the tank and lift

Feeding sticks are great tools to offer your octopus its meal. They can be purchased from aquarium stores, or you can make them yourself from a few simple items.

Feeding Stick Tip

If your octopus always wants to play tug of war with the feeding stick, and sometimes you'd rather not, just slip a plastic drinking straw over the top of the stick and let it slide down to rest on the piece of food. When the octopus grabs the feeding stick, it's really holding onto the straw, and you can slip the feeding stick out. The straw can be quite a plaything and a treasure for an octopus, too.

out rocks to find the octopus. Just have a careful look for clues like exoskeleton remains. Captive octopuses will often have a midden where they dump remains. Always keep an eye on water quality with your test kits.

Once the octopus is used to being fed by you it will associate your presence and the opening of the lid with dinnertime and come to the surface in anticipation of food. Through association with food you will eventually be trusted by the octopus and in some cases met at the surface with arms reaching out, begging for food!

The Feeding Stick

Using a feeding stick can be a very effective means of offering food right to your pet octopus. This tool simply involves spearing an item of food onto the end of a piece of rigid plastic airline (or use two bamboo skewers tied together—they will float to the tank's surface and are easily removed once the octopus has become bored with them) and offering the food straight to the octopus in its hiding place. This is a great way of encouraging a more-settled octopus into taking nonliving food, as you can wiggle the stick about a bit to simulate movement and catch the octopus's interest. After a while, the stick method can turn into a game—a real tug of war between you and your octopus. Don't be surprised if an overzealous octopus makes off with the food *and* the feeding stick!

If you have a female octopus with eggs that is staying in her den, you can use the feeding stick to offer her food. This is an effective way of making sure she has the opportunity to take food, but despite all efforts most brooding females will never feed.

Hand-Feeding an Octopus

Whether to feed an octopus by hand or not is debatable. There is no benefit to the octopus in being fed by hand over the previously described methods, so it is really something that you should carefully think about before doing. The simplest advice is: don't do it.

Yet there are many people who hand-feed octopuses. They enjoy the tactile experience and believe that the octopus benefits from the contact as a source of enrichment. If you are convinced you want to hand-feed an octopus, be aware that you stand a good chance of being bitten and that all octopuses are venomous to an extent. Early cephalopod-keeping information that said species such as *O. bimaculoides* would not bite is wrong! Any time you come into physical contact with an octopus—or a cuttlefish for that matter—you are putting yourself at risk of being bitten or at least nibbled.

The two most common ways people can be bitten are if the octopus mistakes your finger for food and bites you or it can envelop your hand and bite you thinking you are food, especially if you were handling food. You can avoid both by using a feeding stick. There have been no reports of home aquarists (or at least their hands) being pounced on by their octopuses.

As already mentioned, all octopuses are venomous to some extent. All octopuses also are inquisitive animals, and part of the way they explore their environment includes biting to find out what is edible and what isn't. It is best then to always assume that if you are hand feeding an octopus that you are about to be bitten! The result of such a bite will vary depending on the species and the size of the octopus. Symptoms will include a sharp pain from the beak's penetrating your skin and a subsequent reaction from any envenomation that has occurred. You may have localized swellings, shortness of breath, dizzy spells, and nausea. Some people are more at risk than others when it comes to the effects of envenomation and anaphylactic shock. If you are bitten, and if you have an allergic reaction to the octopus's venom, *you must seek emergency medical care immediately*.

The results of hand feeding can be unpleasant in other ways if you have a large octopus. A medium- to large-sized octopus is a very strong animal, which will frequently decide to take the food *and* your hand back to its den. It can often take quite some time to untangle you hand and wrist from an eight-armed wrestler! Suckers can leave their prints, like small bruises, on human skin, and you are likely exposing yourself to a serious bite when you try to pull your arm back.

Taking the opposite view, if you ever put your hands into an octopus tank, or in fact into any aquarium with livestock, you should always ensure that your hands are clean and that you are not adding any pollutants or chemicals into the water. Examples of potentially dangerous substances on your hands include insecticides, nicotine, skin lotions, perfumes, dish soaps, and house-cleaning chemicals. In most cases it is safer, cleaner, and healthier to keep your hands out of the octopus aquarium altogether.

If you must have physical contact with your octopus, there are other ways to do so without involving hand feeding. You could try to touch its arms and even try to gently pet it. Some octopuses will respond to that, especially between and above the eyes. They have even been known to thrust their heads up looking for another head rub.

FEEDING CUTTLEFISHES

Unlike octopuses, cuttlefishes are quite commonly sold as eggs and can be hatched out within the aquarium. For this reason alone, having a good reliable food supply of the right size is of paramount

A Risky Gamble

While we're talking about putting your hands into an octopus tank, we'll remind you that certain species, such as the blue ring octopuses in the genus *Hapalochlaena*, should never be kept—whether or not you ever stick your hand into the aquarium. Keeping blue-ringed octopuses is foolish and extremely dangerous—an aquatic version of Russian roulette, if you will. Although they are not especially aggressive and there are not that many recorded deaths from blue-ringed octopus bites, the venom is deadly and rapidly acting. Why take the chance? Accidents can happen. These animals should never be kept in a home aquarium.

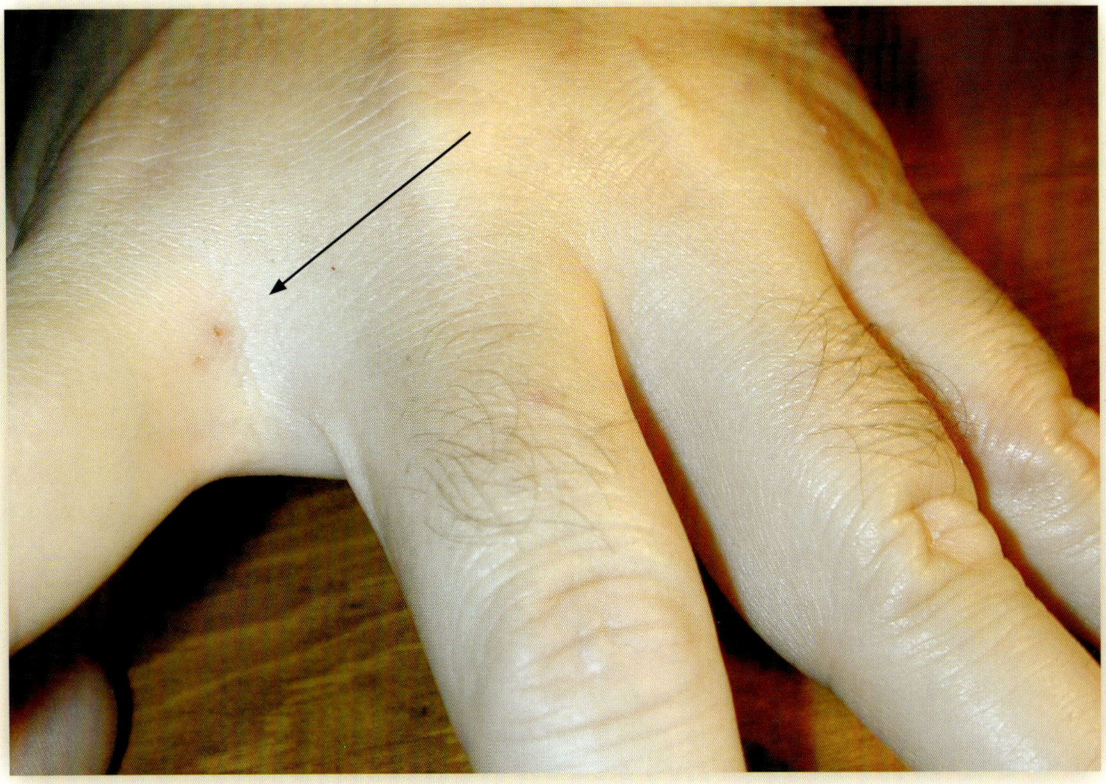

Nancy's husband Bill was bitten by their *Octopus bimaculoides*.

importance. Young *Sepia officinalis* will hatch out at about 10 to 12 mm (two-fifths to one-half of an inch) in total length, and *S. bandensis* will be much smaller at about 6 mm (one-fifth of an inch).

At the newly hatched stage, baby cuttlefishes normally do better if they are housed within a small plastic breeder tank floating within the main aquarium. Make sure that there is always access to live food and lots of it. Baby cuttlefishes will not eat dead food and often will be as old as three months or so before they will—if ever.

If there is not enough food present in the tank the young cuttlefishes will eat each other. A typical *S.*

Harmful Nematodes

Cephalopods can harbor nematode larvae from genera such as *Ascaris* and *Anisakis*, which pose a potential risk to humans. These parasites can be transferred during handling the animal with bare skin.

This cuttlefish is extending its feeding tentacle. In addition to eight arms, cuttlefishes have two feeding tentacles for capturing prey.

officinalis baby of about 2 cm (less than an inch) will eat up to six 2-cm-long shrimps each per day. They eat much more than a comparably sized octopus of the same age.

Cuttlefishes can increase their mass by approximately 25 percent each week if fed correctly, and it won't be long before they can be released into the main part of the aquarium. As they continue to grow you must simply offer larger and larger prey items.

Just like octopuses and their leftover foods, you must ensure that the cuttlefish aquarium is kept clean and free from buildups of carapaces and half-eaten body parts. They are every bit as messy as octopuses and in fact probably more so, as cuttlefishes do not create a midden and seem more prone to catching and killing a moving target just for the sake of it, even when not hungry.

The carapaces left in a cephalopod aquarium can cause ammonia spikes as they decompose and also cause phosphate levels to rise, so these items are best removed periodically to help prevent unwanted algal blooms and toxic conditions.

Another aspect of a cuttlefish aquarium, which can make it messier than an octopus's tank, is that they are more sociable with conspecifics and accordingly can be kept with other cuttlefishes of a similar size. However, more cuttlefishes equals more waste, so keep more than one only if your tank is big enough and your filter system and water-change regimen can cope.

One thing in favor of cuttlefishes and their feeding habits is that they will not prey on hermit crabs or turbo snails or annoy brittle stars to distraction. This means that you can employ those animals' services in much the same manner as in a typical reef aquarium and have a large cleanup crew of well-protected crustaceans, mollusks, and echinoderms who will scrounge around the substrate and eat any leftovers. That said, if you add a bunch of hermit crabs, a cuttlefish will try to eat them but will eventually get bored and wait for easier-to-eat food from you.

Large cuttlefishes need daily feedings to stay in optimal condition, and they do best on a diet of crabs and shrimps. Like octopuses, many of them learn to accept thawed frozen shrimps.

TYPES OF FOOD

Variety is the spice of life, and a healthy balanced diet will give you a healthy octopus or cuttlefish. You should offer different foods from one day to the next.

Crabs and Hermit Crabs

In the wild, most octopuses will feed on a large amount of crabs. It would seem that all crab species are suitable, although some crabs are more aggressive than others and can fight back. *Necora puber* comes to mind and laid claim to a few sore fingers in Colin's early rockpooling years! For this reason it is important to feed your octopus or cuttlefish an appropriately sized crab.

Crabs have ten legs, the front two of which have taken on the form of pincers. They use these pincers to feed themselves, to fight within their species, and to protect themselves from predators. A crab can inflict a fair amount of damage to a cephalopod if given half a chance. It will fight to the bitter end trying to free itself and in the process it will nip and tear at the arms and any other part of the cephalopod's body that comes within reach. They can cut and even sever arms completely.

It is important to make sure that you do not add to the tank any feeder animal that might cause bodily harm to your pet. As mentioned earlier, a good rule of thumb for using crabs as food is to make sure the crab is no larger than about one-third to one-half of the octopus's mantle size.

The collection of crabs can be done in a similar fashion to collecting shrimps, and here, too, make sure that you are not breaking any laws beforehand. One of the best ways of collecting crabs is to visit rocky shorelines and check out rockpools at low tide. By turning rocks you should be able to find some crabs of a suitable size. Always remember to put the rocks back exactly as you found them, as a lot of other animals depend on these rocks as refuge. By leaving the rock upside down you will kill them by exposure to the air or leave them open to predation from sea birds.

Fiddler crabs might be found in the wild, depending on where you live. They are also commonly sold as bait for fishing and are available from a number of online sources. If you decide to use bait animals always keep them in a feeder tank for at least a few days to monitor their health before offering them to a cephalopod. This period of time is good for feeding the crabs, as they may not have been fed for many days and will be of less-than-perfect nutritional value. Give the crabs fish flakes and try small amounts of various kitchen scraps. They are opportunistic scavengers and will try most foods.

Cuttlefish Foods

- *Mysis* shrimps
- other living small shrimps
- frozen shrimps
- crabs

Crabs are one of the best foods you can offer your cephalopod. Even with finicky eaters, crabs almost always seem to draw interest.

In an emergency, crabs can be bought from local fish shops that sell marine invertebrates, or alternatively there are a few species of freshwater crabs commonly sold—the red claw crab, *Perisesarma bidens,* for instance. However, this should always be a last resort, as it is both costly and dangerous if the crab was kept in a tank that previously had a copper-based medication used in it to treat diseased fish.

Fellow hobbyists who keep reef aquariums may be quite happy to give you crabs that sneaked into their aquariums. These hitchhikers can cause problems to corals and other inhabitants if left unchecked. In fact, one of the reasons octopuses were traditionally imported for the aquarium trade in the UK was to be added to a reef tank for a period of time to rid the live rock of any unwanted pests such as crabs and mantis shrimps before the ornamental stock was added. Octopuses are still imported into the UK for this reason.

The crabs most often seen for sale in seafood shops are normally too large for use except with the largest octopuses in captivity. Some shops will sell bags of crab claws or smaller crabs, but you will need to do some research into what is local to you and what your cephalopod will or will not eat. In the U.S., it is almost impossible to buy fresh crab meat or fresh-frozen crabmeat unless you're willing to pay to have someone in a fish market or seafood shop prepare the raw meat for you; almost all of it is pre-cooked before sale.

Crabsticks are not an option for feeding to cephalopods or in fact any other aquatic animal. They are a product specially made for human consumption and contain preservatives and other ingredients not suitable for aquaria.

Crayfishes/Crawfishes

Larger than freshwater shrimps, these lobster-like crustaceans can be located easily in fishing shops as bait. They are eagerly eaten by larger octopuses and cuttlefishes, but some caution should be exercised before offering them. It would be wise to keep them in their own tank for a few days; they may be starved and in poor health when you get them. If you live in an area where you can buy them fresh for human consumption, this is even better than those sold for bait.

Some crayfishes bred for human consumption have grown to the size of small lobsters with formidable claws. Choose females that have smaller claws for feeding to an octopus, and remember that crayfishes should be used only as an occasional treat.

Some crayfish species are sold in aquarium stores for keeping as pets. It would be prohibitively expensive to buy these ornamental crayfishes as food, but just as with the cherry shrimp, if home-bred they could be an additional item on your cephalopod's menu.

Crayfishes should only be offered, if offered at all, as an occasional treat. They are freshwater invertebrates, and freshwater food items do not provide optimal nutrition for marine creatures.

Fishes

There has been a fair amount of experimenting with noncrustacean foods for cephalopods with very mixed results. The use of feeder fishes is, at the present time, still widely acceptable in the U.S., more so than in many other nations, and commonly associated with presenting live goldfish or guppies to large predatory fishes.

From a dietary point of view, the fatty deposits within freshwater fishes tends to suggest that using them as a food item will possibly decrease the life expectancy of a cephalopod. Another very important reason for not using live fishes is in regard to the fish themselves and where they came from. For example, goldfish will normally have been bought from a local fish store, and there is a very strong likelihood that the fish was treated with copper-based medications or at least was swimming in a tank that had been treated with copper. For this reason it is not worth the risk of feeding such a fish to an octopus or cuttlefish, because even small amounts of copper introduced into the cephalopod aquarium can be fatal.

Freshwater Shrimps

There are more and more species of freshwater shrimps making their way into the aquarium hobby these days. They come in a variety of sizes, and species and some of them are even reasonably easy to breed in small aquariums. *Neocaridina* species include the popular cherry, crystal, and bumblebee shrimps, which attain a maximum length of only a few inches and are maintained as algae eaters in planted aquariums. Tests have shown that the *Neocaridina* species will live in marine water for a few hours before dying if uneaten by a cephalopod.

The Amano shrimp from Japan, *Caridina japonica,* is also popular in planted aquariums but has a more complicated life cycle and is therefore harder to breed (the shrimp move to sea water for a planktonic lifestyle when young and return to fresh water as they mature).

Whether freshwater shrimps will make a suitable long-term diet for cephalopods has yet to be established. Unless you have an assembly line for breeding a suitable quantity of them, they will be prohibitively expensive in any case. One of the authors of this book, Colin, has reared over 300 cherry shrimps starting with eight adults in a matter of only three months in an aquarium of 12 x 8 x 8 inches (30 x 20 x 20 cm). It is certainly a route that requires more research, especially in the nutritional enrichment of these shrimps as a suitable food for young cephalopods.

Artemia

This genus of shrimps needs no introduction to people who have been in the hobby for any time at all. They go by the common name brine shrimp and are universally used as one of the first foods that fish fry in breeding projects are offered.

Trouble with Shrimps

Not all octopuses are successful in catching live shrimps, especially when they are younger. Some species, such as *Octopus briareus,* can catch shrimps and even fish by extending an arm and grabbing the prey or by using the webbing between their arms like a parachute to engulf the food item. *O. bimaculoides* has been observed to have a more difficult time during the first encounter with a shrimp. You need to take into account that individual octopuses may vary in their ability as well.

Caridina shrimp.

There have been many attempts by hobbyists and scientists to use *Artemia* as a food source for cephalopods, but in almost all cases the broods of octopus paralarvae die off or fail to grow properly. These crustaceans can be enriched to be more nutritious for feeding to cephalopods but are still thought of as being junk food. Avoid *Artemia* as a cephalopod food.

Marine Shrimps

Probably just about any species of marine shrimp would make an ideal food item for a cephalopod, but several species are more available than others. They can be collected from beaches (laws permitting), they can be bought from fishing tackle shops where they are sold as bait, and they can be sold live from seafood shops for human consumption.

Live or fresh uncooked shrimps are always preferable to cooked shrimp. A raw shrimp is normally a gray color and only after cooking does it turn pink—the color we are more familiar with.

If you are lucky enough to live within traveling distance of a beach, you could collect your own shrimps. There may be local laws that prevent collection in your area, so please check with the authorities first. Many shrimp species are suitable for use; do some research into the species that might live near you.

What About Raw Shrimps?

Many supermarkets these days sell fresh raw shrimps for home cooking, and they are perfect for our uses with cephalopods. Try to avoid buying shrimps that are precooked and pink in color, as they have the lowest amount of nutritional value. They may also have been treated with a preservative, so you should not use them except in absolute emergencies. It's also worth noting that if you are using frozen shrimps they should be thawed fully before being offered to a cephalopod and that they are best thawed at room temperature. Defrosting food in warm water removes some of the nutritional value.

The most commonly used species are from the North Atlantic or the Gulf of Mexico and include the genera *Corophium*, *Crangon*, *Mysis,* and *Palaemon*. All of these shrimps are easy to care for and, luckily, reasonably easy to find. These species are also available by mail order from online shops and can be shipped to you.

Corophium volutator

The smallest of the species we have used is the mud shrimp, *Corophium volutator.* As its common name suggests it lives on the tidal estuarine mudflats. The mud shrimp is a detritivore that actively scavenges through the surface layer of muck looking for food. In the aquarium mud shrimps are easily maintained on a diet of crushed fish flakes. Don't be tempted to recreate the muddy habitat—it can quickly foul your water and block filters. A substrate of washed sand will suffice.

Their average size of 5 to 10 mm (one-fifth to two-fifths of an inch) makes them an ideal first food for newly hatched octopuses and cuttlefishes. *O. briareus*, *S. officinalis,* and *S. bandensis* have all been reared on a diet of *Corophium* shrimps.

If you want to try and catch this or a similar species yourself, be aware that mudflats can be a dangerous environment and that you should seek local knowledge for a safe area. You will need a net, a bucket with a lid, and if the weather requires, some method of cooling the collected shrimp. A great piece of equipment worth investing in too is a battery-operated air pump and its airline and air releaser, because the shrimp require well-oxygenated water during transportation. A cheap way of keeping them cool for transportation during warm weather is to fill a couple of empty plastic bottles with water and place them into the freezer the night before. Make sure you fill them only three-fourths full to allow for expansion as the water turns to ice. Add these bottles to the buckets of water for the return journey. Don't open them, as the melting ice will lower the salinity of the water; instead simply let the bottles float on the surface, which will keep the water cool for a reasonably long time.

Catch the shrimps by using a long-handled net, preferably one with a triangular opening having the base of the triangle at the front. Without digging into the substrate, skim the net across the surface of the mud where the water is only a few inches deep, and every so often stop and rinse the net in sea water, then tip the shrimp into a bucket filled with water from the collecting locality. It might take you some time to find a good collecting spot, but trial and error will suffice. Once you have collected as many as you need, add the bottles of ice, insert the airline and air releaser tightly through the lid of the bucket (the lid will have to be drilled first) and then make sure the lid is sealed. If the weather is really hot; take the fastest route home. If you need a large quantity, take several buckets instead of cramming many shrimps into a small space.

Crangon spp.

Crangon shrimps are great for cephalopods as well. These shrimps, which average 1 to 2 inches (2.5 to 5 cm) in size, prefer a sandy substrate into which they will burrow to find food and can often be seen buried with just their eyes and antennae visible. *Crangon* will prey on both *Corophium* and *Mysis*, so either these species should be kept in a separate tank or you should collect only one species at a time.

Visit the pools left on sandy beaches as the tide goes out to catch these shrimps. Lightly trawl the surface of the sand with your net. This method of shrimping is actually very interesting from our point of view as aquarists, as there are many different species that can be found at the same time. You may find other shrimps, gobies, crabs, and even small flatfishes.

Mysis spp.

Other shrimps that are suitable are those of the genus *Mysis*. They are most frequently encountered in estuaries alongside the mud shrimp, but instead of being benthic they are free-swimming. Their transparent body lends them a ghostly appearance as they swim about. Sometimes they can be seen in large schools in pools of water left by the tide going out—a perfect way to find them for collection. Mysids too are fairly easy to maintain in the home aquarium and can be caught in the same manner as the mud shrimp. They average just under 1 inch (2 cm) in total length and are good for small to medium-

Crangon crangon is an ideal food for all cephalopods.

Mysid shrimps collected in a net.

sized cephalopods. There are many different species of *Mysis* shrimps and sizes may vary, but they can all be encouraged to reproduce in a captive environment.

Palaemon spp.

The shrimps that belong to the genus *Palaemon* are larger than any of the ones mentioned already. There are several species of *Palaemon,* and all have the characteristics and shape most people associate with shrimps. *P. serratus* has been used with success by Colin—this species is a denizen of the intertidal rockpools in the UK, continental Europe, and down to Africa. Other species are found in Florida and are sold via mail order and online in the U.S. They are good as food and also for octopus enrichment, as they are not easy to catch. They are also good scavengers in a cephalopod's aquarium if not eaten right away.

Catching these shrimps can be a bit more time consuming than catching the others. They often must be

Sloppy Eaters

Cephalopods can be really messy when eating. They discard a lot of shells, heads, carapaces, claws, legs, and many other body parts of feeder animals. With octopuses the remains are often dumped into a pile known as their midden. In the wild this offers no problems as the seas are so vast, but in an aquarium it can quickly lead to water quality problems. Always remove any uneaten food as soon as you see it and remove the content of an octopus's midden at least once a week. If you have live rock in your cephalopod tank, then with some luck you will have introduced some amphipods or other scavengers like bristleworms. Amphipods will be eaten by small octopuses and cuttlefishes, but large specimens will totally ignore them. A cephalopod tank will often end up with a large population of amphipods and bristleworms because of the amount of waste from the cephalopod's eating habits, and that is not a bad thing since they will be helping to tidy up.

The Right Foods

Over the years there have been attempts made using different types of food for cephalopods. This is the trickiest part of the hobby for many people, especially those who are landlocked or have no sources of live or frozen crustaceans. Trial and error has shown that the best food for a pet octopus or cuttlefish really is the same as what they would eat in the wild, and it can't be emphasized enough that before you get a pet cephalopod, make sure that you can provide the correct food for it.

individually netted from rockpools, as they are capable of quickly swimming backwards and even straight out of the water in order to escape your best attempts. However, you may get lucky and sweep a few at a time under clumps of seaweed dangling in rockpools. They are an ideal species to go after for medium- to large-sized octopus or cuttlefishes, as they can reach a size of almost 5 inches (12 cm).

These shrimps can also be bought from seafood shops alive or at least fresh. Try your best for live shrimp; that way you can put them in a separate aquarium and can give them a couple of good meals before adding them to the cephalopod tank.

Mollusks

Some octopuses will feed on other mollusks, particularly bivalves like clams, abalones, and to a lesser extent mussels. This group of feeder animals is the most problematic when it comes to keeping them alive for extended periods of time, and if one dies within your feeder tank and goes unnoticed it won't be long before the water quality deteriorates and all occupants will die. For this reason we recommend that any mollusks bought for feeding to cephalopods be bought fresh and used within a day or two of purchase. Alternatively, if your octopus is content with dead food, you can freeze the shellfish on the day of purchase—after giving the shellfish a good feed if possible.

The actual feeding of bivalves to an octopus can be as simple as dropping one into the tank and letting the octopus do what comes naturally. In the case of a cuttlefish, you can open the cockle or mussel yourself and offer only the meat inside to the cuttlefish, but they really can make a mess of their aquarium with this food. Perhaps other mollusks are best suited as emergency-only food items.

Snails

Octopuses will also eat snails, more often small ones when the octopus is young. Many beginner octopus keepers fear they will never be able to keep a cleanup crew with their octopus, and although it is true that you may have to replace some hermit crabs and snails in the beginning, as the octopus grows it will ignore these smaller food items. If you are out collecting some live crabs or shrimps, bring back a few different snails to try out on your octopus too (if you come across them and are allowed to by law, of course).

Squids

Squid was experimented with as a food item by one of the authors of this book, Colin, with cuttlefishes to see whether they would eat this. It is widely found for sale in supermarkets and seafood shops in a fresh uncooked state.

On the one occasion squid was offered the next day the cuttlefish started to cannibalize each other. It would appear that squid and cuttlefish taste quite similar and that one of their normally minor arguments got more serious when they discovered that their fellows taste just like food! Squid was never offered again, and the overly aggressive behavior stopped.

FEEDER TANK

If you are planning to set up a tank for an octopus or any other cephalopod, you might end up needing to set up at least one extra tank—one tank for your cephalopod and one tank for its food!

As we've stated before, cephalopods, especially newly acquired or juvenile ones, need live foods to trigger a feeding response. In some cases you may get lucky and an individual will recognize nonmoving items or unrecognizable shapes of food as being edible. However, especially in the case of young cuttlefishes, if actively moving live food is not used, they often will not feed.

Remember that in the wild the octopuses and cuttlefishes that we are likely to encounter in the aquarium trade will prey upon crabs, shrimps, and sometimes other cephalopods. The backbone of their captive diet should always be crustacean based.

Mysis shrimps in a separate feeder tank. They can be collected from such a tank and offered to your cephalopod when needed.

This photo illustrates a simple setup for maintaining fiddler crabs. All that is required is damp sand and a small "pond" of brackish water. Note the strawberry and lettuce for food.

The Right Size

The feeder aquarium has to be functional in the sense that it will be very easy to maintain while keeping the feeder animals alive as well as being suitable for the animals themselves. It must be large enough for maintaining a few days' or weeks' worth of food for your cephalopod. Normally a tank of about 15 gallons (60 liters) is suitable for keeping 50 to 60 small crabs or 100 shrimps. This can provide food for a young octopus for a few weeks.

The Right Setup

A typical feeder tank set up to house marine shrimps, crabs, or bivalves should have a thin layer of sand, preferably coral sand or aragonite. These substrates act as a pH and hardness buffer and with the addition of a few rock caves or plastic pipes, it will provide the animals with some form of security. The sand (and the tank for that matter) should be well washed before use and cleaned on a regular basis with a gravel cleaner. The depth of the substrate is very important from the point of view of maintenance, as most crabs, shrimps, and bivalves that are used as feeders like to burrow and will happily do this in the feeder tank. Unfortunately, the feeders may not always be in the best of health when you receive them and there will be some fatalities. If they have burrowed into the substrate and have subsequently died, the water quality will deteriorate rapidly; if action isn't taken quickly the all of the animals in the tank may die. Keep a close eye on the feeder tank and look after your octopus's food carefully. Ironically, to be a successful cephalopod keeper you will firstly need to be a successful crustacean and bivalve keeper.

The shrimp aquarium will essentially be just like a normal aquarium but smaller, such as a 10- or 15-gallon (40- or 60-liter) tank. However, if you are setting up a feeder tank for crabs there are a couple of options open to you. If the crabs are from rockpools, offer them hiding places in the form of clay plant pots, plastic pipes, and other caves similar to those provided for the cephalopods. The crabs also appreciate being able to leave the water for periods of time, so you should allow some of the rocks to emerge from the water. The water in the shore crab tank needs to be only a few inches deep. There have been occasions where live crabs have survived a water quality crash in their tank by climbing out onto rocks to escape the toxic conditions, so having these perches is especially recommended.

Staples of the cephalopod diet: shrimps, crabs, and clams. These can all be collected, brought home, and maintained in a feeder tank until it's time to offer them to your cephalopod.

Observe how this cuttlefish's colors change when eating. In the picture on the left you will notice how the cuttlefish is muted, almost the same color as the tank's substrate, as it slowly approaches the offered food. In the picture on the right, the colors change dramatically as the cuttlefish grabs its meal. Stimulating your cephalopod's natural hunting abilities and techniques is an excellent way to keep it active and healthy.

Keeping fiddler crabs is easier with regard to water quality—these crabs have a more amphibious lifestyle. In the wild they will generally live in mangrove or seagrass areas which have fluctuating temperatures and salinity levels. They spend a great deal of time picking over the mud for food, or they might live on the sand near the water, hiding among grasses or other vegetation.

In the home, you can keep them in a tank or container mostly consisting of damp sand with a bowl serving as a pool of water. The water in the bowl should be brackish, made up of half salt water and half RO/DI water which you can mix in advance and keep in a jug. You should replace the water in the bowl and dampen the sand with this water every day. The crabs will burrow in the sand and use the bowl several times a day to keep damp. These crabs can be kept indoors at room temperature.

In addition to fish flakes sprinkled on the sand, fiddlers enjoy the occasional piece of air-popped popcorn (no butter!) and a bit of ripe fruit or lettuce. Be sure to remove the uneaten food promptly to keep the tank clean.

Your cephalopod can learn to take dead food items, but usually this takes time.

Feeder Diet

The healthier you can keep the feeder animals the better it will be for the animals destined to eat them. One of the most important issues is that the feeder animals must be fed well themselves. A feeder animal is more nutritious the better it has been fed.

Before putting any live food item into the cephalopod tank it is advisable to feed the crabs or shrimps a good-quality fish flake food about an hour before you need them. In the case of bivalves, which are filter feeders, they do well being fed on liquid feeds as sold for fish fry and supplemented with cultured algae sold as phytoplankton for reef tanks.

Feeder Tank Maintenance

Due to the fact that the feeder tank never has a stable population of animals, its water quality is often in a state of turmoil and ultimately prone to swings in ammonia concentration and pH. The filter should therefore be above specifications for that particular volume of water. A protein skimmer and the use of carbon filtration are also required to maintain the best conditions. Water changes should be done on a regular basis, and test kits are essential. Pay particular attention to ammonia, nitrite, and pH.

DEAD FOODS

Newly hatched and juvenile cuttlefishes and octopuses require live foods. You will be able to witness the true nature of the cephalopods while they are hunting, changing colors and skin texture to outwit and ambush a shrimp. However, with time and a little bit of training you should be able to get your cephalopod onto nonliving foods. If you can get your pet cephalopod to eat these food items, feeding time suddenly becomes a much easier option in that you can collect your own crabs and shrimps and freeze them for later use. Make sure that all food is thoroughly defrosted before it is added to the aquarium. Having said that, it is always better to use well-fed live crabs and shrimps at every opportunity you can.

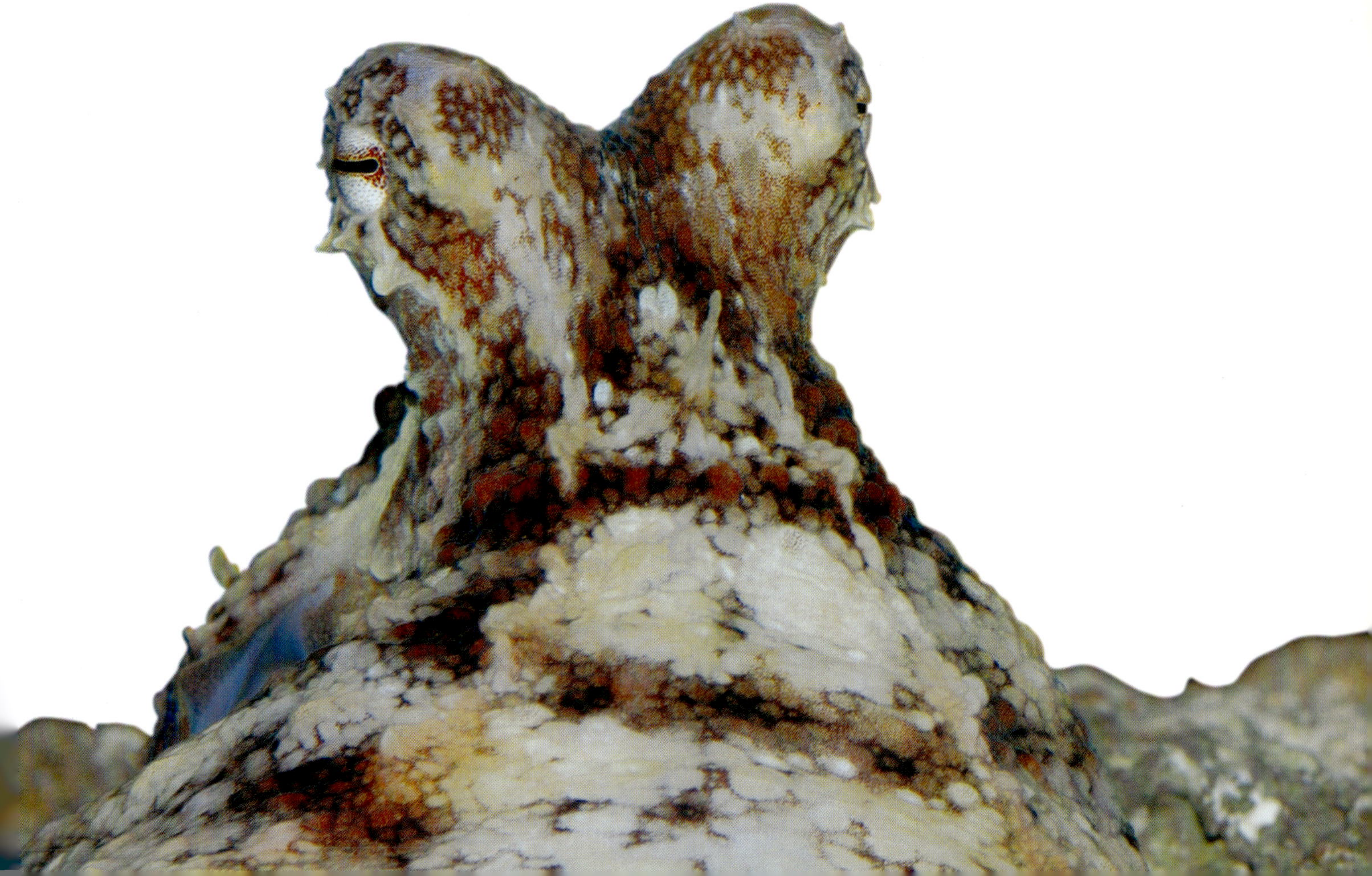

Chapter 7

Breeding Cephalopods

Here is undoubtedly the newest and most exciting part of cephalopod husbandry: breeding and reproduction. There can be no better goal to set and certainly no better benchmark to achieve to show that your pet cephalopod is healthy than its willingness to breed.

FROM an ecological perspective, it makes sense to try to maintain a sustainable captive-bred stock of cephalopods. At the time of this writing, almost all cephalopods being kept in captivity are wild-caught, and only a small percentage of these wild-caught animals were collected from the wild as eggs and hatched in captivity. A huge number of the collected eggs would have succumbed to predation in the wild, so collecting eggs should not damage wild stocks. (For a brief overview of cephalopod reproduction, see Chapter 1.)

SEXING YOUR CEPHALOPOD

If you wish to breed octopuses and acquire them as hatchlings or juveniles, most likely you won't be able to determine the sex, so you'll need at least several of them. Male octopuses are usually recognized by a specialized third right arm known as the hectocotylus and may have larger suckers near the base of their second and third right arms. The distinguishing characteristics of male octopuses become more visible as they reach maturity, but still, it may be difficult to sex the octopuses while they are going about their business in your tank. Some, such as *Octopus bimaculoides*, are notoriously difficult to sex under the best of conditions. Home aquarists usually adopt a wait-and-see policy.

Sexing individual cuttlefishes for a breeding project is not usually as critical as it is with octopuses. This is because cuttlefishes are generally quite willing to live as a group and if there are at least four to six of them

Eggs produced by an *Octopus briareus*. You can already see tiny octopuses growing inside some of them.

in an aquarium the chances of getting a compatible pair breeding is quite high. In the most commonly kept species, such as *Sepia officinalis*, the males generally grow larger, display frequently to the females once sexually mature, and just like octopuses have a hectocotylus. However, with cuttlefishes it's the fourth arm which is modified for breeding, not the third. Sexing cuttlefishes is not always reliable until they are at or near sexual maturity—initially the disparity between sizes is the first clue.

OBTAINING EGGS

There are three ways to get octopus eggs: breeding a pair of captive octopuses, buying eggs, or having a female octopus that encountered a male before coming to live with you lay eggs while you have her. There have been rare occasions where octopus eggs have been offered for sale, but this really is unusual. Perhaps if you live near someone who is breeding large-egg octopuses or has a female that has laid eggs, you might be able to obtain a festoon of eggs and try to rear them to hatching, but at the time of this writing this is very uncommon.

By contrast, cuttlefish eggs are becoming quite commonplace on the lists of marine fish exporters and can be found for sale in the U.S. and the UK without too much hassle. If you live elsewhere, ask at your local aquarium store and see what they can do for a special order. Ask them to look for cuttlefish eggs—or "squid eggs" as they are sometimes (incorrectly) called—and see whether they can check the scientific names. The biggest problem is with the species you are ordering and the reality that the seller might not actually know what species it is that they are exporting! Having said that, the most commonly seen species being exported as eggs is *S. bandensis*, and their rearing is fairly straightforward.

Cuttlefish eggs are normally sold by the bunch and look like black grapes. An average exported clutch will contain about 15 individual eggs, and the hatch rate is usually quite high. It is often possible to see the developing embryos through the egg wall if they are lit from behind. Just as with any other aquatic animal, you should acclimate eggs slowly to the new tank's water parameters.

Some clumps of eggs never hatch. Perhaps they were infertile or damaged in some way. If you are there in person to choose your eggs, take the ones that are swollen or, even better, you might be able to see ones with cuttlefishes inside them.

S. officinalis can also be purchased as eggs, but this species is less common in the U.S. It is not found along any part of the U.S. coastlines—as related earlier in the book, no species of cuttlefish is. *S. officinalis* is found in UK and continental European waters. In the UK there are some companies that specialize in selling marine life to public aquariums, and they can often obtain European cuttlefish eggs.

EQUIPMENT, TIME, AND EXPENSE INVOLVED

If you intend to breed cephalopods you are going to have to spend a good deal of time and money in setting up your facility. There is a need for all of the same equipment as is required for setting up a normal cephalopod aquarium, but you are going to need a lot more of it.

You will need at least two tanks for your breeding stock—your adult octopuses or cuttlefishes. Two tanks are required in case you need to split up the cuttlefish pair because of fighting, which could happen if the pair is not compatible, or perhaps, in the case of some species, you can only put the adults together for breeding and immediately have to split them up again.

You are also going to have to consider what to do with all of the baby cephalopods when they hatch out. The female octopus will care for the eggs, but as soon as the larvae emerge her care will normally end. A female cuttlefish will show no interest in her eggs after she has laid them. This makes removal of the eggs a bit easier once she has finished spawning and you have taken over the care for the young. You also could remove the parents and leave the eggs where they were laid.

With octopuses you will normally have to catch the young and put them into a different tank for closer, and therefore easier, observation. This will be no easy task, as they may hatch over several days and you will need to be on duty constantly. After a short period on the glass (using their suckers to hang onto the inside panes of aquarium glass), the tiny octopuses disappear into the tank's live rock, so you need to capture them before they hide. A floating breeding net (or one attached to the tank's rim) is often used, but a separate tank might be easier. Either way you are going to require more space than one aquarium to do this right.

You might want to consider setting up several tanks all plumbed into one filter. This is commonly done by using a sump for all the tanks in your system. This means that as water leaves each tank via an overflow

Octopus mercatoris hatchlings fighting hours after hatching.

it all ends up in the same sump and is treated by the same protein skimmer and other filtration methods. You will need to ensure that the filtration unit and skimmer are large enough to cope with your tanks' entire volume of water.

If you have utilized this particular method, adding a new tank to the system is fairly straightforward and can often be done with immediate results, as the water in the system is already mature.

Breeding cephalopods will consume quite a bit of your time. You will need to be on hand during octopus introductions and matings just in case things take a turn for the worse and a fight ensues. And looking after what could potentially be a few hundred baby octopuses or cuttlefishes will certainly eat away at your free time. There's water to be tested and changed and lots of hungry mouths to feed!

Feeding the young is the hardest and potentially the most expensive part of the job of breeding cephalopods. So far only large-egg species of octopus have been reared by anyone outside research institutes. Small-egg species have almost impossible requirements for zooplankton. In any event, the sheer volume of small live foods needed for the large-egg species is enough to make the most hardened of aquarists weep!

There are a lucky few who breed cephalopods and live close enough to the sea in which they can catch their own live foods by the bucketful and bring that home to their babies. For everyone else the fact is that unless they are breeding amphipods or *Mysis* shrimps at home, they are going to spend a lot of time and a lot of money buying suitable small live foods.

For a bit of perspective; a female *O. briareus* might lay 100 eggs, and they might all hatch into young octopuses with an arm spread of approximately 10 mm (two-fifths of an inch). Some babies will die within a few days, and this seems to be quite common regardless of how well they are fed. But even if we assume that you are trying to rear 30 newly hatched babies, each individual will need four or more amphipods per day—that's 120 amphipods a day for all 30! And that's only to begin with, because as they grow they eat more and more. Within a couple of weeks you'll be going through about 2,000 amphipods a week, and then you will need to move onto bigger and more expensive food items.

You've been warned! Check the price of buying this food and having it shipped to you. To reduce the number of young octopuses you may have to feed, consider giving some of them away to other cephalopod fanciers. Although the odds are against it, hatchlings have been known to survive without special care in a mature reef tank with plenty of existing food like amphipods. You can also research how to culture your own live foods.

MY OCTOPUS JUST LAID EGGS...

If you have a mature female octopus, it is more than likely that sooner or later she will lay eggs. This can catch a new octopus keeper totally by surprise. Many commonly kept species are difficult to sex—*O. bimaculoides*, for example—and you may think you have a male, only to be surprised by eggs one day. The fact is that female octopuses can arrive in your tank already carrying fertile eggs because they mated with a male before collection. Or they may simply store the sperm until they are ready to produce the eggs.

Whether or not you can hope to rear these eggs into baby octopuses depends on whether they are from a

An *Abdopus aculeatus* female brooding eggs, keeping her arm over them. Unfortunately, small eggs are almost impossible to raise in captivity.

small-egg species (next to impossible to care for when they hatch) and whether you have the space, time, and money to care for large-egg species when they hatch. In any case, the female octopus has probably made a den for herself and has secretly laid her eggs on the roof of the cave and they are hanging down in clumps called festoons. She is a very attentive mother and will be constantly running her arms across the eggs keeping them free of detritus. Your once-friendly pet octopus may now act very aggressively if you attempt to go near the den and will very often flash warning colors.

At this stage there is nothing you can do except to leave the female be with her brood and let her do what comes naturally. You can try to feed her using a feeding stick, but even if she attacks the food and takes it from you she might not eat it. Her action could have been a defensive maneuver rather than a feeding response, so don't assume she has eaten. Check for uneaten food. A female often will not eat once she has laid eggs, but she might at least for a while.

Large or Small Octopus Eggs

As you no doubt have already inferred, octopuses are lumped unscientifically into two groups as far as egg-laying is concerned: large-egg and small-egg species.

Small-egg species will lay literally thousands of tiny eggs that will hatch out into paralarvae that require a stage of their development to be floating around with other zooplankton in the oceans and feeding on microscopic fauna. This has proven to be a very tough situation to replicate in aquariums.

Large-egg octopus species have no planktonic stage, and when they hatch they are fully formed small octopuses that are almost immediately benthic in their lifestyle and can be successfully reared in the home aquarium by offering small animals such as *Mysis* shrimps and amphipods.

Time Required For a Hatch

The incubation period of an octopus's eggs will vary because of two main factors. From one species to the next there will be a variation in the number of days it takes from being laid until hatching time. (You will have a clue that the hatching time is getting close when the un-hatched octopuses can be seen within their egg cases and they will flip over 180° in the egg.) The temperature of the aquarium's water will be the other main influence on the length of incubation. The warmer the water, the quicker the eggs will hatch out. This can provide an advantage for us if we are faced with possibility of rearing some young octopuses, because we could split the batch and keep them at two different temperatures. However, removing any eggs from a protective, and probably angry, female octopus will be a task in and of itself; she will try to protect all of her brood, but if possible you could remove some eggs and hatch them in a separate aquarium, at a higher temperature of a few degrees warmer than the female's tank. The advantage is that you will have two chances at hatching eggs and looking after young octopuses. Treat the first batch to hatch as a practice for the second batch. That way you are literally not keeping all your eggs in one basket.

Removing a female octopus's eggs may cause her to hold a grudge, and because she can tell people apart,

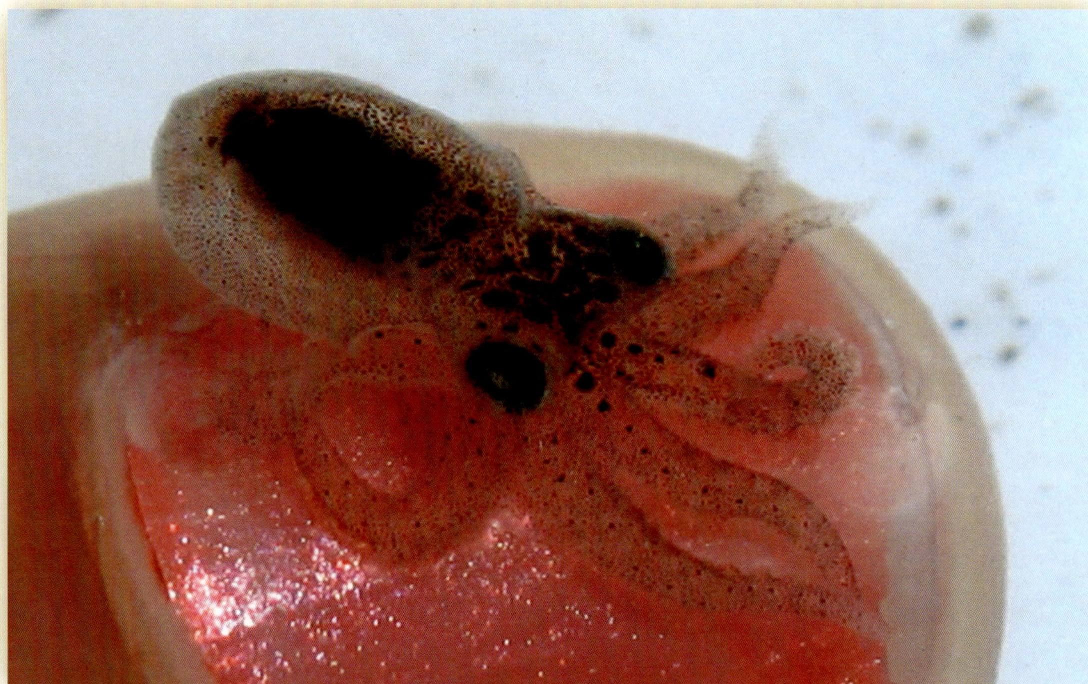

A three-day-old *Octopus bimaculoides* on a human fingernail.

Abdopus aculeatus **pair, male (left) and female (right), introduced for a short period to observe mating.**

you may find your once-friendly octopus irritated and seemingly angry when you're around, but being friendly to everyone else.

The Eggs in an Aquarium

Female octopuses will be constantly cleaning the eggs and ensuring that there is sufficient water flow around them to supply oxygen to the unhatched cephalopods. If you remove some eggs from the female you should place an airstone nearby and try to suspend them from something like a plastic clothes peg to allow them to hang down, the way they were laid in the female's cave. Not all species attach their eggs to a substrate, however. Some will actually carry them about, so this advice may need to be modified depending on which species you have.

MY CUTTLEFISH JUST LAID EGGS...

First of all, give yourself a round of applause for finding them! A female cuttlefish will try her best to hide the eggs she lays in small crevices away from potential predators. They are dark in color, and she will try to secrete them in the corners of the tank where it is not as bright.

The eggs are best removed from the breeding tank and reared in a separate tank or breeder net suspended in the main tank.

A one-month-old *Octopus bimaculoides*.

Cuttlefish eggs can be kept in a separate container inside your main aquarium. You will need to move the hatchlings to the main tank once they are too large for the smaller tank within a tank.

Role of the Cuttlefish Mother

In contrast to the role of the female octopus, a cuttlefish's commitment to her offspring ends when the eggs are laid. Some hobbyists who have bred *S. bandensis* have mentioned that the female cuttlefish spent several days mating and laying eggs and hanging about in the same area as the eggs were being deposited. We still don't know whether this constitutes any parental care or not. Time and research and more breeding attempts will tell.

Time Required for a Hatch

The two most commonly cultured cuttlefish species are *S. bandensis* and *S. officinalis*. They have very different requirements as far as their temperature ranges are concerned (see Chapter 8), so there are differences in their incubation. At their preferred temperatures in the high 70s, the tropical *S. bandensis* will hatch after three or four weeks. The temperate *S. officinalis* will hatch in the same amount of time but at a suitable temperature in the 60s.

What Percentage Will Live?

In the wild most of the newly hatched cephalopods will perish. This is a fact of life, and the two different strategies with regard to eggs size work in different ways. The small-egg species can release as many as half a million paralarvae into the seas with the hope that a few survive long enough to grow and eventually

A container holding a very young *S. bandensis* has been placed over a dime to illustrate the size of the tiny cuttlefish.

breed. Large-egg octopus species give their young a head start and skip the planktonic stage, allowing them to immediately act like miniatures of the adults instead of floating about in the ocean's currents. In contrast, the large eggs-species may lay only tens to hundreds of eggs.

If you provide the right conditions, can provide enough suitable food, and have enough space, you can expect a great many of the young ones to survive. You may actually want to leave some of the young octopuses in with the adult and have some in separate rearing tanks. Just remember that even in a tank without predators the young octopuses can be preyed upon by siblings.

Aquarium Predation

In the aquarium there are no predators, so that is one factor removed from the survival equation… or is it? Are you using live rock? There might be bristleworms in there that are large enough to overpower vulnerable newly hatched octopuses. Likewise, remove serpent stars, as they can easily catch and kill young cephalopods.

HOW CAN I FIND HOMES FOR MY NEW CEPHALOPODS?

Re-homing your surplus offspring should prove to be quite easy. There are always people looking for cephalopods, and the fact that yours are captive bred is a great advantage. You can post advertisements on Internet forums and inquire at your local fish stores as to whether they are interested in buying any. Most young cephalopods should be at a suitable size for re-homing when they are approximately two to three months old.

Chapter 8

Species Notes

This chapter includes descriptions and notes on the care of the species commonly kept in home aquariums, plus some species that you may want to keep but shouldn't. In this chapter we will review octopuses and cuttlefishes, as well as argonauts, nautiluses, and squids. For a refresher on setting up a tank for a nautilus or a bobtail squid, see Chapter 3.

IDENTIFYING OCTOPUSES

Octopuses are notoriously difficult to identify. They are creatures of change: arms can be stretched out to look longer or contracted to look dumpy, and colors, textures, and shapes can all be modified. An octopus may easily resemble another species. You may buy an octopus and never be able to fully determine what species it is—this is fairly common. Luckily, though, octopus husbandry is not too different from one species to another.

Nevertheless, it is important to try to determine where it came from. Insist that your local aquarium store or online supplier go back to the person who provided them with the octopus to learn more about where it was found. This may give you some indication as to what species you have, and it will also tell you about the required water temperature.

Take plenty of good photographs for determining species and post them on websites like www.tonmo.com to ask other hobbyists for a possible ID. Also note whether your octopus is nocturnal (you may not know this until it has had some time to settle down), has webbing between its arms, has an ocellus (false eye spot), or is heavily textured. Note what colors you commonly see. Photos will provide more information about body shape, relative arm length, and generally what it looks like.

OCTOPUSES

Abdopus aculeatus

Abdopus aculeatus is a small octopus found in shallow tropical waters from the Philippines to Indonesia to the Great Barrier Reef in Australia. *A. aculeatus* occurs in a tropical intertidal habitat, on areas of coral, rubble, and sand. In its natural setting, it forages at low tide for small prey such as crabs.

A. aculeatus is a small yet very long-armed octopus, with a mantle about 2 inches (5 cm) long and arms reaching up to 12 inches (30 cm). It often has a grayish-black-ocher appearance when found in its ocean habitat, and it appears to keep at least some of this coloring in the home tank.

This species has remarkable camouflage ability. It can appear very textured, even raising its skin to make it look like weeds and seaweed, giving the octopus protection as it forages during the day. Recently this species was featured in the news as the "walking octopus." Wearing what looks like a seaweed disguise, *A. aculeatus* was filmed walking backwards on two arms. You might see some of this behavior, or at least the seaweed imitations, if you keep it in a tank large enough for it to display its natural defense behaviors.

A. aculeatus is a diurnal octopus, which is a big advantage for you as an octopus keeper. For this reason, and its relatively small size, it is imported and kept by a number of home aquarists in the U.S. and is possibly the most commonly imported species available in the UK.

In addition to being a good candidate for your aquarium, this species has demonstrated friendly behavior towards both keeper and carefully selected tankmates and shows no particular problems. Although *A. aculeatus* is far-traveling in its natural range, it can live comfortably in a 50-gallon (200-liter) tank at tropical temperatures (78°F, 26°C).

A sand substrate would be appropriate, as well as live rock and low-light, nonstinging corals. Because this

This *Abdopus aculeatus* mimics seaweed as a defense and is moving along by walking on two arms.

species has been seen gathering pebbles to put in front of its den, you might want to add a few pebbles and shells to the tank.

Like all the other octopuses in this genus, *A. aculeatus* has arms of varying length, with the front arms being the shortest. In addition to camouflage or disguise, it has another interesting defense: its arms are capable of being severed at the base, so if attacked by a predator an arm breaking off provides a means of escape. The octopus sacrifices an arm to get away, just as many lizards may shed a tail to escape a predator. (This can also happen if the octopus becomes stressed during handling.) As in other octopus species, the lost arm will grow back.

Other Abdopus *Octopuses*

There are several very closely related species within the genus *Abdopus* that could be easily confused with *A. aculeatus*. Unfortunately it would be difficult for a hobbyist to distinguish among them without the species descriptions and access to a scientific laboratory. For this reason "*A. aculeatus*" is a bit of a catch-all name at present within the hobby, though care for all of the closely related species will be the same.

A mating pair of *Abdopus aculeatus*.

Mating is carried out with the male and female a long distance apart, with the male stretching his hectocotylus (third right arm), modified for reproduction purposes, over to the female. *A. aculeatus* produces small eggs that hatch into planktonic young that are extremely difficult to raise, which is a real pity. Were it a large-egg species, with easier-to-rear young, it could be a perfect species for captive propagation and supply to the home aquarium hobby.

All *A. aculeatus* specimens and the related species you find on sale will be wild-caught. At the time of writing this book, this species is not threatened or endangered in the wild.

Enteroctopus dofleini

Enteroctopus dofleini, the giant Pacific octopus (GPO), grows too large to keep in all but the biggest home aquariums. It is included here because every octopus keeper sooner or later will aspire to keeping one. The attraction of having a big, intelligent, usually friendly octopus that can live up to five years is irresistible. However, the reality is that this much-admired species looks cramped in even an 800-gallon (3,200-liter) tank and needs chilled water (50°F, 10°C). Even with a tank of this size they are usually kept in a flow-through system in public aquariums. They eat a proportionally large amount of big crabs daily, so therefore huge filtration and equally huge protein skimming is required.

Found in the Pacific Ocean along the coasts of northern California, Canada, and Alaska and all the way to Japan, the GPO lives at a range of depths, from shallow waters along the shore to deep waters more than 4,900 feet (1.5 km) down.

The GPO is a large, handsome octopus, typically red-brown at rest, and often has very textured skin in the form of creases or flaps. It is nocturnal in its habits and at night feeds on typical octopus prey such as crabs, lobsters, shellfish (often clams, which it drills with its radula), and various other foods. They have even taken food as unusual as seabirds and shark eggs. There is video footage of a GPO catching and eating a shark in a public aquarium.

Because of its large size and long lifespan (up to five years), the GPO is popular in public aquariums around the world. An adult can have arms over 6 feet (1.8 m) long and a body which can easily reach 2 feet (60 cm) in length. There are reports of much larger GPOs, but these reports are unsubstantiated.

E. dofleini moves out into deeper waters to mate and then returns to shallower waters, where the female lays her eggs. Like most other male octopuses the male has a long modified arm tip that it uses during mating to pass a 3-foot-long (1-m-long) spermataphore (sperm package) to the female. The female lays up to 100,000 small eggs that take six months to hatch into paralarvae that join the rest of the ocean plankton until they grow larger and become benthic.

The giant Pacific octopus.

The Blue-Rings I Have Known

One of the authors of this book, Colin, has kept blue-ringed octopuses on two separate occasions, and both were by accident. It may sound strange to talk about keeping an animal accidentally, but these octopuses were exported from point of origin under false identities. In 2002 he ordered two *Octopus vulgaris* from a tropical fish importer. Colin had been searching for this species for quite some time, but when they eventually arrived at the fish importer's facility, the transport box was opened to reveal two tiny *Hapalochlaena* octopuses nestling in separate bags between a clownfish and damselfish shipment. Luckily Colin was there to collect the octopuses immediately. We dread to think what might have happened if the staff had been in the position where they had to acclimate the octopuses alone with no idea of what species they were dealing with! As it turned out, one of the blue-ringed octopuses was dead in its bag and the other was dead by the next morning. Blue rings ship poorly.

Fast forward to 2008 and it happened again, only this time on a far greater scale. While working on this manuscript we required some species of octopus for observation and photo opportunities. Colin had a rare opportunity to obtain six *O. chierchiae.* A day before the octopuses were due in the UK an update e-mail arrived from the supplier saying that the exporter didn't have any *O. chierchiae* and had sent substitutes. (This is a common practice with exporters who make up the value of the order because shipping live fishes and invertebrates is quite costly, since they are charged by the box. The box should be full to make the process as cost-effective as possible.) Unfortunately, and much to the surprise of both Colin and the shopkeepers who ordered the octopuses, the exporter had substituted 28 blue-ringed octopuses! This led to a flurry of phone calls to friends who worked in public aquariums to find homes for these venomous octopuses.

When they arrived in the UK, 21 of the 28 octopuses were still alive in their bags. Over the next 24 hours and after very careful acclimation more died, and within a couple of days only five were still alive. Eventually only one survived. More proof, if any was needed, that their survival rate is exceptionally poor both during and after shipping and that these cephalopods are best left in the oceans—not just because of how lethal they are, but also because they have a terrible survival record.

Giant Playtime

Giant Pacific octopuses usually react favorably to being given toys and respond to various other enrichment techniques. Aquarium visitors like to watch these large octopuses engaged in activities such as playing with children's toys.

Perhaps GPOs are mostly associated with Puget Sound, a portion of the Pacific Ocean in northwestern Washington State where divers take an octopus census every year. Most published photos of divers encountering large octopuses feature the GPO.

The expense of acquiring a GPO (the cost of the animal plus freight) pushes this species out of range for the casual cephalopod keeper. It is best left to the public aquariums and experienced marine aquarists who win the lottery!

Hapalochlaena spp.

There are a half dozen described species in this genus, with more to be described. All of them are unsuitable for the home aquarium. *Hapalochlaena lunulata,* the greater blue-ringed octopus, is the most commonly imported blue ring, though not intentionally. It usually arrives at the wholesaler as a substitute for another species.

This is a small octopus with a mantle length of less than 2 inches (5 cm) and arms normally less than 3 inches (7.5 cm). It is a tropical octopus, found on coral reefs and rubble where it hunts for crustaceans and other animals to eat. *H. lunulata* is found in the western Pacific including the Philippines, Indonesia, and Papua-New Guinea. Its iridescent blue rings are flashed as a warning when the octopus is provoked. Otherwise this octopus is good at camouflaging itself against its background.

Keeping blue-ringed octopuses has always been a subject of great debate. Many online discussions about this species have quickly deteriorated into heated arguments covering many aspects, from the

Blue Ring Venom

Blue-ringed octopuses are infamous for their venomous nature. Their bright blue markings, which are so typical of species in this group, serve as an important warning that the saliva produced by these animals contains a powerful neurotoxin called tetrodotoxin (TTX). If a person is bitten by a blue-ringed octopus, the injected TTX acts as a nerve blocker and will prevent messages from the brain from getting to voluntary muscles, leaving the victim paralyzed. Involuntary muscles like the heart will continue working normally, and a person can remain fully conscious through the whole ordeal. Survivors report that they remember everything that happened despite their completely unresponsive state. Victims who survive have received immediate medical care, including breathing support on the way to the hospital.

Blue-ringed octopuses are absolutely stunning creatures but also are potentially deadly. Their venom can kill. It is best to leave this beautiful cephalopod in the ocean.

responsibilities that come with keeping a highly venomous animal to a person's right to keep a deadly animal if they want to. Let's review a few facts about blue rings.

- The bites from blue-ringed octopuses can kill.
- In some cases a person might not even realize they have been envenomated, because often no pain is felt. By the time dizziness and weakness have taken hold it might be too late. If you get bitten, assume you have been envenomated and seek immediate medical attention.
- This genus of octopus ships poorly, and many experts believe that this is partly because they are often flushed from their dens in the wild by use of chemicals. At least a third of all exports die in transit or soon after.
- Blue-rings can have a natural lifespan—from egg to death—of only six to eight months, so they are very short lived in aquaria, especially since they normally arrive as adults.
- Experts all agree that octopuses of this group are best left in the sea or kept only by public aquariums and laboratories with experienced staff.

Octopus bimaculoides

Octopus bimaculoides, commonly referred to as the bimac in the hobby, is one of the most popular octopuses to keep in the home aquarium in the United States. Native to the mid to southern coast of California, its range extends down to Mexico's Baja California peninsula. Bimacs can occasionally be found in tidepools but may live in water as deep as 65 feet (20 m) in mud, sand, and on rocky reefs. It is sometimes found in or near the same area as *O. bimaculatus*, a close relative but identified by a different ocellus and the size of the eggs it lays.

A bimac is a small to medium-sized octopus requiring at least a 50-gallon (200-liter) tank. On average, its mantle will be about 5 inches (13 cm) long and its arms up to 14 inches (36 cm) long. Sometimes known as the California two-spot octopus, it has an ocellus with an unbroken blue chain on a black background, as opposed to the broken chain on *O. bimaculatus*.

This is Ollie, a bimac belonging to Nancy. Ollie is displaying a purple color on her arms. Octopuses can change both color and texture seemingly at will.

Bimacs are very personable and will interact with you.

This ring becomes iridescent during hunting and at certain other times, depending on mood. The pattern of the unbroken blue chain is unique to this species.

Some Californians actually collect their own bimacs from tidepools and shallow water. (If you plan to do this, check to see what sort of permit is required.) *O. bimaculoides* has been the most common captive-bred species of octopus in the home aquarium trade but is available as a wild-caught specimen. Breeders sometimes have been known to sell bimacs online, but what is listed as a bimac does not always turn out to be that species. You should thoroughly research your purchase and even ask for photos.

Bimacs are good octopuses to keep in home aquariums for many reasons. They are large enough to be relatively easy to rear from eggs, but they still stay small enough to fit in a 50- to 75-gallon (200- to 300-liter) tank. Although not as heavily textured and not as colorful as some other species, *O. bimaculoides* has the advantage of being diurnal, so it will be active during daylight when you are around to observe it. They are usually friendly—or learn to be after getting over their initial shyness—and will spend a great deal of time in plain view. Many of them will interact with their owners and enjoy physical contact such as being petted between and above the eyes. They might give you a curious nip when they are young, but their bite is small and not harmful, provided you are not allergic to the venom. (Seek immediate medical attention if you are.)

A bimac changes color and texture, but not as dramatically as some other species. At rest it has a mottled gray pattern with some yellow. Some people think this looks like a floral pattern. Bimacs seem to depend on their color more than skin texture when they conceal themselves, but like other octopuses they are very good at camouflage and hiding. When younger and less confident, they hide on or between rocks, so it takes an experienced eye to spot them.

Unlike most other octopuses available for a home aquarium, bimacs come from temperate water rather than from a tropical zone. This means you would rarely need a heater in the tank, as room temperature is normally adequate; in fact, a common concern is that the tank's temperature does not climb too high. A bimac would be happy at 65°F (18°C) or even lower, but it can live at room temperature (72°F, 22°C). Although individuals of this species have been kept in water with a temperature in the mid 70s, there is some indication that higher temperatures may shorten the lifespan because of quickened maturity. Given that a water pump in the sump and even dim lights will raise the tank temperature, many bimac keepers resort to a chiller or a fan on the sump to regulate the heat. Another advantage of the bimac is that it is not as likely to escape as certain other species—*O. vulgaris* for example. However, some will crawl out of the tank, so you still must seal it like the way we recommend for all octopuses. (See Chapter 4.)

In home aquariums, bimacs can live over a year, but this is dependent on what age the octopus was when you received it. In this respect, a captive-bred specimen with known age is a better purchase.

It is very difficult to determine the sex of *O. bimaculoides*, so much so that even home aquarists who have kept a number of bimacs cannot determine whether they have a male or female. This is because the reproductive arm of the male (third arm on the right) does not look much different from the corresponding arm of the female.

Once sexually mature, a female *O. bimaculoides* finds or builds a cave that then becomes her den. A female will lay up to 800 eggs ranging from about one-fourth to three-fourths of an inch (about 6 mm to 2 cm) in size. The eggs will hatch in four to six weeks, depending on tank temperature (in a warmer tank they will hatch sooner). As the eggs are laid, they are attached in festoons to the wall or top of the den. The female bimac tends the eggs, blowing them with her siphon to keep them clean and further cleaning them with her suckers. She guards the eggs and will try to fight off any intruders, including you if you attempt to take her eggs. She may not eat during this time, but food should be offered. She will probably not live long after the eggs hatch.

Bimacs are a large-egg species, which means that their hatchlings are fully-formed small bottom-dwelling octopuses capable of finding food items (copepods and amphipods) in a mature tank right from the beginning. The fact they don't have a paralarval stage is why they can be easily cultured in aquariums. They are often started on mysids and then move to the larger foods.

If a female is in good condition she will lay eggs as an adult whether she has been mated with a male or not. An unmated female will go through all the same processes as building a den and laying the eggs, but the eggs are infertile and will not hatch out. You might get an indication that this is happening if your previously interacting octopus refuses to come out of her den, even for food.

Bimacs are sometimes raised from eggs in batches of similar-sized siblings. If there is not a suitable den

Octopus briareus are sometimes kept in a carefully constructed reef tank. Note the deep webbing of this young *O. briareus*.

for each of the young octopuses there may be some cannibalism, so supply each one with a small shell or piece of plastic pipe for refuge. The ones that survive seem to get along well.

Moving a pair of them to a home aquarium and expecting them to live together has not been successful in the many examples we have seen. One rises to dominate the other and often the subordinate octopus doesn't survive. Don't plan to keep more than one bimac in a single aquarium.

Provided you are experienced with keeping a saltwater tank, a bimac makes a great first octopus or a good choice for an experienced keeper who wants to rise to the challenge of breeding a cephalopod or raising this species from eggs.

Octopus briareus

Octopus briareus is a long-armed octopus found in the Gulf of Mexico, the Caribbean Sea, and the northern part of eastern South America. It can be found on coral reefs and in shallow waters. A nocturnal octopus, it hides during the day in holes in the reef and comes out at night to hunt. *O. briareus* feeds by using the deep webbing between its arms to cover corals or rocks like a large umbrella. Then it flushes out small prey into the webbing using its long arms (it catches mostly crustaceans and small fishes).

O. briareus is a small to medium-sized octopus, with a mantle about 4 inches (10 cm) long and arms as long as 24 inches (60 cm). It can be recognized by its long arms, very thin webbing between the arms, bulging eyes, and sometimes a bluish-green iridescent color. Not all of its arms are of equal length; the second and third pairs of arms are the longest. You may observe a brown and white coloration, the body a red-brown and the arms white.

This species is available sometimes and can be successfully kept in at least 50 gallons (200 liters), but given its long arms, a 75-gallon (300-liter) tank with a reef-like landscape is better. It needs tropical water temperatures of about 78°F (26°C). Although nocturnal, *O. briareus* often adjusts to a daytime schedule and comes out in the light. It is a friendly octopus, extending its long arms to touch the glass where your hand is. It's also a beautiful octopus.

It might be tempting to put this octopus in an established reef tank, but this is not a good idea. Its delicate skin can be harmed by stinging corals, and in an enclosed environment like a reef tank, the octopus cannot easily avoid such corals. It will eat some of the reef inhabitants, but others will be a danger to your octopus. *O. briareus*, like the other species discussed in this book, needs a tank of its own.

O. briareus is a large-egg species, so its young can be raised by the home aquarist if given the right conditions. Both authors have had the opportunity to raise newly hatched young ones, which were fed on mysids and amphipods from the beginning.

This species is not as easy to find as some of the other popular species described here, but some Florida collectors will occasionally offer them. Many individuals are caught and exported as large adult females that are already gravid and will lay eggs in the aquarium. This should be a candidate species for captive breeding programs.

Like *O. bimaculoides* and other octopuses of its size, *O. briareus* can live for more than a year under good conditions in a home aquarium.

Octopus mercatoris

This is a dwarf (pygmy) species found in the Gulf of Mexico and the Caribbean Sea; most specimens on sale in the U.S. come from the Florida area. It has become available to hobbyists in the last few years as a by-product of the live rock industry in that state.

Octopus mercatoris is generally reclusive and nocturnal, living under rocks at depths of less than 30 feet (9 m). Not much is known about this species's natural history, but it is often confused with another species found in the same regions—*O. joubini,* the small-egg Caribbean pygmy octopus. One main difference lies in the size of their eggs and larvae. *O. mercatoris* produces large eggs and larvae and *O. joubini* lays small eggs that produce planktonic larvae. Only *O. mercatoris* can be raised by the home aquarist.

A Good Species to Keep

By adapting to its nocturnal habits and observing them at night using only red light, home aquarists have found *Octopus mercatoris* to be an interesting species. On a positive note it seems to be one of the few species that can live in small groups in a suitably big tank when introduced together into an aquarium as hatchlings. Some *O. mercatoris* keepers have even raised several generations of this little octopus in their tanks with success.

O. mercatoris has a mantle size of up to about 1.75 inches (4.4 cm) and proportionally short arms between 2 and 3 inches (5 and 7.5 cm). It is often reddish-brown in color but sometimes shows a blue-green iridescence when young. It has minor cryptic pattern ability, ranging from very smooth to displaying a small "horn" over each eye and scattered bumpy peaks on the mantle.

Coming from tropical waters, it can be kept at 78°F (26°C) and does not require a large tank: 30 gallons (120 liters) is ideal for a first-time octopus keeper, while a 20-gallon (80-liter) tank is acceptable for experienced cephalopod keepers. Add barnacles and other shells to serve as homes and smaller shells to serve as den doors. Octopus-proofing the aquarium is less stringent than for larger species and requires little more than lowering the water level and making sure that equipment in the tank is protected with mesh or sponges so the small octopuses cannot reach in or escape.

Until recently, *O. mercatoris* was normally only kept by home aquarists who found it hitchhiking on Florida live rock. Not all home aquarists who discovered the little octopus in their tanks were exactly pleased. Given its appetite for crustaceans, it can make a considerable dent in the inhabitants of a reef tank by eating the cleanup crew and any ornamental shrimps. If you do keep this species, feed it a crustacean diet, mostly small crabs and shrimps. It will usually learn to take food from a feeding stick or pipette.

In the last few years, more octopus keepers have been choosing *O. mercatoris*, or perhaps they buy it under the impression that it is another species. It has been sold as the "brown octopus" or even more incorrectly as "*Octopus vulgaris*."

Octopus mercatoris.

Despite being normally reclusive, some *O. mercatoris* kept in daily frequented areas of the home will actually interact with their owners. Many will come out and explore and eat while you're watching. This is a big step forward in keeping *O. mercatoris*. It's not for everyone, but if you keep a late-night schedule and want to take the time and effort to work with this nocturnal species, you will be rewarded.

Because it is a large-egg species, it's possible to raise the new hatchlings through to adulthood. It may be the easiest octopus to work with if you want to try your hand at captive breeding and at raising hatchlings. It is a good starter octopus if you don't mind its nocturnal habits, and it's the only readily available octopus for someone able to have only a 20- or 30-gallon (80- or 120-liter) tank.

Octopus vulgaris and Other Large Octopuses

The name *Octopus vulgaris* is often assigned erroneously by local fish stores and online shops. It is the default name given to just about every brown octopus or to any species that confounds the seller. Many home aquarists purchase what they think is *O. vulgaris* only to discover they have bought a totally different species, sometimes a fully grown dwarf. However, this might actually be a good thing, because *O. vulgaris* is a large octopus requiring a very large tank (150 to 200 gallons/600 to 800 liters minimum), and few home aquarists are set up for such an octopus with respect to tank size or food supply.

The *O. vulgaris* species complex comprises more than ten closely related species found in tropical and

Octopus vulgaris.

Octopus cyanea.

temperate waters over the entire world. The ones that are sold in the United States usually come from Florida to Haiti (the Caribbean common octopus). If you live in another part of the world, the *O. vulgaris* (or *O. vulgaris* look-alike) you find for sale would be from near your own area.

Often referred to as the common octopus, *O. vulgaris* can live in a variety of habitats (rock, coral reefs, sand, or mud), may or may not have an ocellus, and may be diurnal, nocturnal, or both. Its front arms are the shortest, with longer second to fourth arm pairs.

The Caribbean common octopus has a mantle length of around 8 inches (20 cm) and arms up to 31 inches (80 cm). It is found on coral reefs, sand, or even sea grass in shallow water and feeds on crustaceans, shellfish, and fish. The common octopus from the UK and Mediterranean area is a bit larger, and the Japanese version is a bit smaller.

Females produce small eggs that hatch into very small planktonic hatchlings known as paralarvae that are very difficult to rear, even in a lab. There have been very few successful attempts.

This is a friendly octopus that people enjoy working with. The common octopus is also well known as an escape artist, so the tank should be well sealed. It has the advantage of a slightly longer life span, usually living up to a year and a half to two years.

If you keep this species, make sure you know which area of the ocean it came from and whether it is from tropical, subtropical, or temperate waters so you know what temperature to keep your tank.

From time to time other large octopuses come on the market. An example is *O. cyanea*, usually from Hawaii, but this species is also found on coral reefs in the Indian and western Pacific oceans. Another name for *O. cyanea* is the day octopus or the Hawaiian day octopus because it is diurnal. It's a bit smaller than the Caribbean common octopus, with a mantle of about 6 inches (16 cm) but arms of the same length, 31 inches (80 cm). The day octopus lives on coral reefs and forages using the same technique as *O. briareus*, enclosing corals and rocks with its webbing and then using the arm tips to flush out prey (the usual crabs, other crustaceans, and fishes). It has a black false eye spot that's not always visible and rows of white spots on its reddish-brown arms.

Courtship has been observed in the species, and mating occurs from a distance. Like *O. vulgaris, O. cyanea* is a small-egged octopus whose eggs hatch into planktonic paralarvae.

Zebra Octopuses: *Wunderpus photogenicus, Thaumoctopus mimicus,* and Introducing *Octopus chierchiae*

The wunderpus, *Wunderpus photogenicus,* and mimic octopus, *Thaumoctopus mimicus,* are similar not

It is easy to confuse one species of octopus with another. This *Wunderpus photogenicus* can be mistaken for *Thaumoctopus mimicus* by cephalopod hobbyists.

only in appearance but also in their natural history, habitat, and their bad responses to capture and life in captivity. They are both fairly new to science, not being described until the late 1990s. Very little is known about their life cycle so far. Quite often a wunderpus will be offered for sale as a mimic, but this is usually an incorrect identification, whether by design or accident, on the part of the exporter or shop.

A burrowing wunderpus. These octopuses should only be kept by the most experienced cephalopod hobbyists, if kept at all.

The very first thing that both authors would like to point out about these two species is that in our experience they should be left in the ocean. Their behavior makes them fascinating, and their striking colors have enchanted many casual observers, but as pets they make lousy aquarium inhabitants.

Octopus chierchiae is being mentioned here because it also gets stuck with the zebra octopus title within the aquarium trade, but that's about all it has in common with the wunderpus and the mimic. At the time of writing there was quite a stir because some live specimens of this species had just appeared after they were feared extinct in the wild.

Many wunderpuses and mimics are exported each year from traders based in Indonesia and the surrounding areas of Southeast Asia. Sometimes they are specifically ordered, but more often than not they are sent as substitutes within larger orders and end up being a surprise for the importer. Aquarium importers and shops are slowly starting to see that there is no market for an overly expensive animal that doesn't deliver the behavior it is famous for and only lives a few weeks or months in captivity. With time we hope that this filters back to the collectors and they stop catching them altogether.

How much feedback do the collectors of such animals get? If an animal lives only two to three weeks in captivity, do its collectors ever hear about it? Importers who receive these animals as substitutes should make sure that they voice their concerns to the collectors.

The collection techniques involved in the capture of these species are frequently to blame, and it is suggested that they are still being flushed from their burrows with chemicals much in the same way that fishes were commonly flushed from coral heads with cyanide-based liquids or bleach. The people who collect in the area often know the location of several specimens and can wipe out all the mimics and

A mimic octopus showing patterns photographed in the wild.

wunderpuses in a large area in a couple of days.

These octopuses are rare in the wild, with divers often having to spend weeks or months to locate individuals to show to tourist divers—remember that they have only recently become known to science. Worse still, the environment in which these two species are found is considered to be under threat. The muddy substrate in which they live is overexploited as a fishing ground by trawlers, and many areas are being reclaimed from the sea for land for farming and housing.

This leads us to discuss the habitat in which these animals live and the next-to-impossible task of recreating it in an aquarium. These octopuses do not live on rocky reefs. They burrow into the mud, and from these dens they go off searching for food. In an aquarium the mud would constantly be in suspension through the octopus's activity, and this would cause havoc for a filtration system, not to mention the visibility within the aquarium.

Some experienced hobbyists have tried creating aquariums for a wunderpus that allowed the animal to dig into a muck-like substrate held together by macroalgae, and although much better than subjecting the animals to a reef-type aquarium, it is still not ideal, and even with this special treatment they still do not live very long.

The Mimic

T. mimicus gets its common name because of its alleged ability to mimic other sea creatures. It is claimed that this species will pretend to be a poisonous flatfish, a venomous lionfish or sea snake, and perhaps even an anemone or a jellyfish. However, is this really mimicry or something else at work? For instance, other species of octopus will swim in the same fashion as the poisonous flatfish: are they mimicking the flatfish or is that just the easiest way for an octopus with long dangly arms to swim? Perhaps these displays are courtship and part of a sexual display. Perhaps the mimic octopus is simply advertising the fact that it is poisonous or venomous and making a big show of its potential form of self defense? At this point we do not yet have all the answers, but one thing is for sure: this species has not displayed these fascinating traits in captivity. There are even reports of mimic octopuses in the wild and in captivity showing none of the barred coloration and appearing either tan in color or a washed-out gray over the entire surface area.

They have been known to play dead in captivity by turning onto the mantle, exposing the underside of their arms. Is this behavior similar to that of other animals such as some snakes that feign death, or is the octopus actually exposing its mouthparts for a much better chance of biting its attacker to inject venom? This is a defensive behavior seen in other octopuses, so it may not be unique to mimics at all. Photographs show *O. mercatoris* doing this as well as *Hapalochlaena* spp. More work needs to be done on this species to verify the claims about its mimicry and its potentially venomous nature. Is it possible that people so far have seen what they want to see and jumped to mimicry conclusions? In any case, many of the photos and videos of the supposed mimicry are far from compelling.

If you do somehow end up being in charge of setting up an aquarium for a mimic, follow the advice given for other tropical octopus species but bear in mind that this is a burrowing, muck-living species that will require a substrate suitable for its natural behavior. Include a few caves or pipes just in case it wants to make use of them. Crabs make the ideal food and should be offered on a regular basis.

Octopus chierchiae

This attractive little octopus was feared to be extinct in the wild because no specimens had been seen for many years. It is another species that has bold zebra-like patterns across the body and arms, and again this could be a sign that the octopus is venomous or poisonous.

Its range traditionally was around Panama; now it seems as though Nicaragua is also part of the range, as

Leave Them in the Oceans

Television documentaries have made the mimic octopus world famous, and many people with an interest in marine aquaria would like to keep one. However, they are far from being the ideal pet cephalopod. In captivity the mimic octopus has shown no signs of mimicry, and compared to other octopus species it has a very poor survival rate.

Octopus chierchiae.

some specimens have started to appear in the aquarium trade via shippers in this area.

Not very much is known about their captive husbandry at this stage, since they have literally been imported only within six weeks of this manuscript's completion. There is still very much to learn about them, but it is worth mentioning them here in case they become more commonly offered for sale.

There are some important issues for this species that stand out as needing further investigation. As exciting as that sounds with regard to their potential, there is still the unknown issue of their population size. Is this a rare octopus in the wild? Perhaps this is another species whose collection should be discouraged. Hopefully more work will be done with this species to determine its state in the wild.

The Wunderpus

W. photogenicus has one of the best scientific names ever, and it really is a stunning small octopus. It has eyes that are generally raised on stalks longer than those on the mimic, and its coloration is somewhat different—reddish-brown with creamy white markings, whereas the mimic has a lighter base color with dark stripes.

Both *T. mimicus* and *W. photogenicus* lay small eggs and so far have not been reared in captivity. Most of the females who have laid eggs arrived gravid or laid infertile eggs. The sexes can be differentiated by looking at the male's third arm, which will be much shorter than all the others, whereas a female will have

A wunderpus, an octopus species best left to live its life in the ocean.

eight arms of similar length.

Normally these two octopus species grow to have a mantle length of about 2 inches (5 or 6 cm) and an arm length of about 12 inches (30 cm).

In closing, the mimic and the wunderpus—and *O. chierchiae* for that matter—may or may not be rare. There is not enough known about their wild populations, but what is certain is that they have a poor survival rate in shipping, and captive husbandry still poses many issues and problems. Their purchase is not to be taken lightly, and although it may be tempting to rescue a mimic or wunderpus for sale in a shop, perhaps the untimely death of the animal in the shop would discourage sellers from importing one again.

CUTTLEFISHES

Sepia bandensis

There is a species of cuttlefish becoming more and more common in the cephalopod hobby that is entirely suitable for life in a captive environment. Compared to other imported species of the genus it is a dwarf and almost classed as easy to keep and breed. It has several common names in the trade, but to avoid further confusion we will just refer to it by its scientific name: *Sepia bandensis*.

Originally this small cuttlefish was sporadically imported from the Indo-Pacific as adults, and they were notorious for their poor survival rates. Quite often they would die a day or two after importation or arrive in the destination countries already dead in their travel bags. When they did survive shipping they commanded high prices in the shops, often easily over 100 dollars. The fact that they were adults with most of their lives already lived made them an uneconomical choice for a cephalopod aquarium even when they arrive healthy. In fact they were best avoided and considered okay to purchase only if they had survived in the shop for a week or so, which the majority of them didn't.

Specimens that survived the importation process seemed to do quite well once they got established and could be housed comfortably as individuals in a mature 20- to 30-gallon (80- to 120-liter) aquarium. Later on this species became more widely available thanks to success in captive breeding. Eggs and juveniles were sold by home aquarists keeping these cephalopods. As hobbyists had more and more success with keeping this species, they have noticed that *S. bandensis* could have a life span longer than the six months originally predicted. In fact, they will live up to a year, sometimes even longer. It's also interesting to note that they grew larger than thought, reaching up to 6 inches (15 cm) in length, three times larger than expected for wild individuals in the ocean.

Young cuttlefishes, like this *Sepia bandensis*, require large regular offerings of live foods.

In contrast to the early days, this species is now more commonly being imported as eggs and hatched out in captivity with very high survival rates. This is the preferred method of obtaining them. Since the eggs normally arrive looking like a bunch of black grapes, it would appear that the collectors, who ship mainly from Bali, are able to find and identify the eggs in the wild with a degree of confidence and regularity, because the eggs are appearing quite frequently on trade lists. It would also appear that sending this species in egg form is a much better method of shipping, with almost all eggs arriving safely and subsequently having a high hatching rate.

S. bandensis eggs will generally hatch within two to three weeks of being laid, depending on temperature, which should be kept at tropical levels between 75° and 82°F (24° and 28°C). When the eggs hatch the baby cuttlefishes emerge at a size of about 4 to 5 mm (about one-fifth of an inch) in length and will eventually reach an average size of 4 inches (10 cm) and sometimes the aforementioned 6 inches (15 cm). Young *S. bandensis* do best if kept in breeding nets within larger aquariums until they reach about 20 mm (four-fifths on an inch) long. Once large enough they can be released into the tank. Individuals can be batch-reared when they are young, but as they reach sexual maturity it is possibly a better idea to house the cuttlefish singly or in groups in a very large tank. As we stated earlier, cuttlefishes require copious amounts of live foods. You can eventually try to wean them onto nonliving foods.

A one-month-old *S. bandensis*. This cuttlefish is being held in a container within the main tank until it grows large enough to navigate the main tank safely.

If a pair of *S. bandensis* is to be used for breeding, they are best introduced to each other once they are mature. Mating takes place almost immediately if they are compatible, but if they're not a fight will break out and could cost the life of one or both individuals. Make sure that you are going to be free to keep an eye on the pair for the time they are together.

The seemingly reluctant attitude of this species towards sharing an aquarium with conspecifics should be kept in mind when purchasing a batch of eggs for rearing, because the eggs often come in bunches of 15 to 40. Can you find homes for the surplus cuttlefishes when the time comes?

The couple mates in a head-to-head position in which the male's hectocotylus is inserted inside the female's mantle and the spermatophores are passed over to fertilize the eggs. The female then lays the eggs among rocks in clumps that are initially pointed in shape but that later fill out. At this stage they can be removed from her tank to rear separately. The females show no parental care, but it has been noted that some females will hang around in the general vicinity of the laid eggs for a day or so.

Sepia officinalis

The European cuttlefish, *Sepia officinalis*, is quite widespread in its distribution and ranges from the UK to South Africa. It is a temperate-water, large-growing cephalopod species that requires a good deal of space

An Ideal Cephalopod

Sepia bandensis really could be the ideal pet cuttlefish. It is small, easily housed, long-lived (it has a life span that could be over a year), available as an egg, and if given extra attention can be bred in captivity.

A Walking Cuttlefish?

One behavioral trait that is very interesting to watch is the technique *Sepia bandensis* has developed for "walking" across the aquarium floor and when climbing live rock. The cuttlefish uses two of its arms and two lobes on the mantle to creep about and sneak up on prey. You should leave open spaces on the sand for such activities and as burrowing sites. This would mimic its original habitat: it lives in shallow water on sand and coral reefs.

in the home aquarium to be kept successfully. It is not very commonly seen at the present time for sale in the U.S., although it has been in the past, but is one of the easier species to track down in the UK and continental Europe. On average, a wild *S. officinalis* should be able to reach a total length close to 18 inches (45 cm). This must always be on your mind if you are deciding to keep even just one in captivity. They can often live close to two years in the wild, but warmer indoor temperatures can speed up their lifecycle to as little as 10 to 12 months.

Your first consideration should be about the size of the aquarium you can manage to get for keeping this species of cuttlefish. We have always maintained through experience that they should be housed in nothing less than a tank with a capacity of 200 gallons (800 liters). This size of aquarium would ensure enough space for a pair to grow to their large adult size, but once again, bigger is always better. Also, you want a tank with lots of surface area—one that is long and wide is ideal over one that is tall and narrow.

You need to be aware that *S. officinalis* is not a tropical species and will not fare well in an aquarium with temperatures above 70° to 72°F (21° to 22° C). An ideal temperature for day to day husbandry would be in the mid 60s. At warmer temperatures this cuttlefish often becomes unsettled and nervous. They are much more likely to ink at warmer temperatures, and if one individual inks it can set off a chain reaction with others following suit—they can easily black out an entire tank in a few seconds. For this reason a huge filter with an equally huge protein skimmer is definitely required. The reaction chamber on the skimmer on Colin's *S. officinalis* tank was about 4 feet (1.2 m) tall but still would take up to two days to remove the ink from the water!

Since this species too can often be purchased as eggs, this is probably the easiest way to get them—and

Sepia officinalis.

Sepia officinalis will observe you as much as you observe it.

the least expensive, as it cuts down on large volumes of water, which are expensive to ship. The eggs can be hatched in a breeding net suspended in a tank after being acclimated slowly to the new water. It is important at this stage to keep them away from any animals, including the tank's cleanup crew, that may try to prey on them. Once the eggs hatch, the babies can be reared for a few days in the net and then released into the aquarium. The chances are that you will not see very much of them if you have a lot of rocks, as they will spend a great deal of time hiding. Ideally you will have a layer of fine sand on the bottom of the aquarium for burrowing. They will quite often bury themselves by stirring up the sand with rapid movements and allowing it to settle on their backs.

They grow very quickly, sometimes by as much as 25 percent of their mass each week. (Remember to offer lots of live food!) They will be sexually mature at an age of about six to eight months in captivity. The males are overall much larger than the females and often take part in displays to each other and minor scraps. You may see bite marks inflicted on each other, especially on the arms and the back of the mantle. These bites rarely seem to cause any problems. Cannibalism can occur, especially in juveniles when two are

Cuttlefish Injury

In the confines of a smaller aquarium the European cuttlefish is prone to injury and the condition misleadingly called "butt burn." As mentioned earlier, butt burn is caused when the cuttlefish gets a fright and shoots off backwards and hits the aquarium glass. The cuttlebone can become exposed, which leads to infection. Perhaps the ideal captive scenario would be to house them in large acrylic tanks with gradually sloped and rounded edges.

A Personable Cuttlefish

Once they settle in, *Sepia officinalis* individuals are not particularly skittish and respond quite well to their owner as regards recognizing feeding time and being at the front of the aquarium watching all that goes on in the house... even to the point of tormenting the dog with jets of water.

fighting over one shrimp and the winning cuttlefish keeps hold of the shrimp *and* the other cuttlefish and just keeps on eating! Another example of cannibalism occurred when a female who was either not sexually mature or simply unwilling to mate was killed and partially eaten by an overzealous male.

Successful matings can take place over several days in an aquarium with artificial moonlight set to mimic the natural waxing and waning of the moon. A female can deposit up to several dozen eggs, and she will attach them to the various substrates in the tank.

Less Frequently Seen Cuttlefishes

Aside from *S. bandensis* and *S. officinalis* there are a few other species of cuttlefish that may appear in the aquarium trade from time to time. There is still a lot to learn about imported cuttlefishes. It is a fact that positive identification without scientific experience can be nearly impossible. However, what has been noted in the home care of species of cuttlefishes so far is that as long as the basics for the animals are taken into consideration (size and temperature requirement) their husbandry is very similar.

A *Metasepia* species, one of the so-called flamboyant cuttlefishes.

Some imported species will come with an accurate identification, and that is partly because they really could not be mistaken for any other species, such as members of the genus *Metasepia*, also known as flamboyant cuttlefishes. Flamboyant cuttlefishes are offered for sale from time to time and have been kept successfully by home aquarists.

M. pfefferi, is occasionally imported and finds its way to aquarium stores. It comes from the Indo-Pacific region, with its range extending to the northern Australian coast. The flamboyant cuttlefish lives in shallow waters on sand or on mud, is diurnal, and eats crustaceans and small fishes. It has a body length of about 3 inches (7.5 cm). Its dramatic coloration has given *M. pfefferi* its common name. This species shows striking patterns of yellow, maroon, brown, and even red. Like other flamboyant cuttlefishes, this species has an unusual method of locomotion in that it walks along the sand using its lower arms and flaps along its body. It's possible that this cuttlefish might be poisonous, suggested by the remarkably strong coloration.

Being small and not swimming vigorously, this species can be kept in a 30-gallon (120-liter) tank with a sand substrate, with water at a tropical temperature.

ARGONAUTS, NAUTILUSES, AND SQUIDS

The cephalopods we will cover in this section are the ones much less likely to be kept in a home aquarium. They are difficult to find and a challenge to keep, with some of them being almost impossible. We are including them, though, because we feel a book on cephalopods would be incomplete without them. Here we will describe the argonaut as well as certain species of nautiluses and squids that, although not recommended for home aquariums, occasionally find their way into a home tank.

Before we say anything further about these cephalopods, we must voice a strong word of caution: most are better off left in the sea. If they are to be kept in captivity at all, they should be kept only in a marine lab or a public aquarium.

Nautiluses are the uncommon cephalopods most often seen for sale. In certain areas of the world you can find and take home bobtail squids from the sea. Argonauts may be found stranded on the shoreline or swimming the waters near the shore.

Argonauts

The argonaut is the least common cephalopod discussed in this book. It is included here because people have occasionally found them either in the open water while boating or along the shoreline and once in a great while will try to keep them. You will almost never find an argonaut for sale in the aquarium trade. If you do come across an argonaut, it is best to give it to a local public aquarium or research lab or better yet, leave it in the ocean. Argonauts are difficult to keep and tend to live only a short time in captivity.

Named after the Argonauts in Greek mythology, the argonaut has been known since ancient times. It is an open-water cephalopod, actually a type of pelagic octopus, found in tropical and sub-tropical waters

On Buying Unnamed Species

There always will be risks when buying unnamed species. Two important variables that you would not know when purchasing an unnamed species are their preferred temperature and their eventual adult size.

around the world including the Pacific, Atlantic, and Indian oceans. The size varies between species, with the largest shells being up to 18 inches (46 cm), and most species are distinguished by the size, shape, and color of the shell. Males are only about 10 percent of the size of females, usually measuring less than 1 inch (2.5 cm). Argonauts are capable of displaying many colors, even blues and reds.

The female argonaut uses two modified arms to secrete a thin shell that superficially resembles a nautilus's shell but is actually an egg case. Because of their thin shells, argonauts are sometimes referred to as the paper nautilus, but they are a type of octopus. This thin shell-like structure is where the female lays her eggs; the shell helps with buoyancy. As she floats along, a female argonaut keeps her arms inside the shell, with webs covering the outside. The male does not create a shell. Its hectocotylus (reproductive arm) has a pouch that stores sperm. When he encounters a female, the arm is detached and transfers to the female—it is capable of autonomous movement. The male dies soon after mating as in other cephalopods, but the female continues to live and grow and reproduces repeatedly.

Argonaut feasting on a fish.

The argonaut is also a food item to a limited extent and is found in fish markets in India and Japan.

Being open-water cephalopods, even small argonauts require a huge tank of a minimum of 100 gallons (400 liters) to be housed comfortably. Argonauts don't have much control when swimming and, like nautiluses and squids, can run into the sides of the tank. Like nautiluses, they need dim lighting.

Feeding them can be difficult, but you should try small fishes. They rarely eat live food in captivity, and the most success has been feeding them small dead fishes, perhaps even several a day. You may have to animate the dead food to elicit a feeding response.

If you find an argonaut with eggs, there is little chance that you can raise the young, as they will be planktonic.

The argonaut *A. pacifica.*

Nautiluses

Nautiluses (sometimes called pearly nautiluses to distinguish them from the paper nautilus, the argonaut), of which there are several species, are living fossils, not much different from the early cephalopods we find in the fossil record. Today they are found in the Indo-Pacific region on reefs, specifically the outer edge of the reef. They migrate up and down the water column, from 230 to 1,640 feet (70 to 500 m), rising closer to the water's surface at night.

They are mainly scavengers but may also be opportunistic predators. Although once thought to have poorer eyesight than most predators, recently published research suggests that they can see much better than previously thought. Like the other cephalopods discussed in this book, nautiluses eats crustaceans. They in turn are preyed upon by fishes and also by octopuses. Although they are consumed locally as food, nautiluses are most valued by humans for their shells.

Yes, nautiluses have shells—one of the characteristics they share with their ancestors. The shell consists of calcium carbonate, which the nautilus secretes as it grows. It's the only cephalopod living today with a complete external shell, although remnants of the shell exist in other species.

Most of us are familiar with the cross section of a nautilus shell with its many chambers spiraling out from a center point. The nautilus itself lives in the first chamber, the chamber closest to the opening. It can

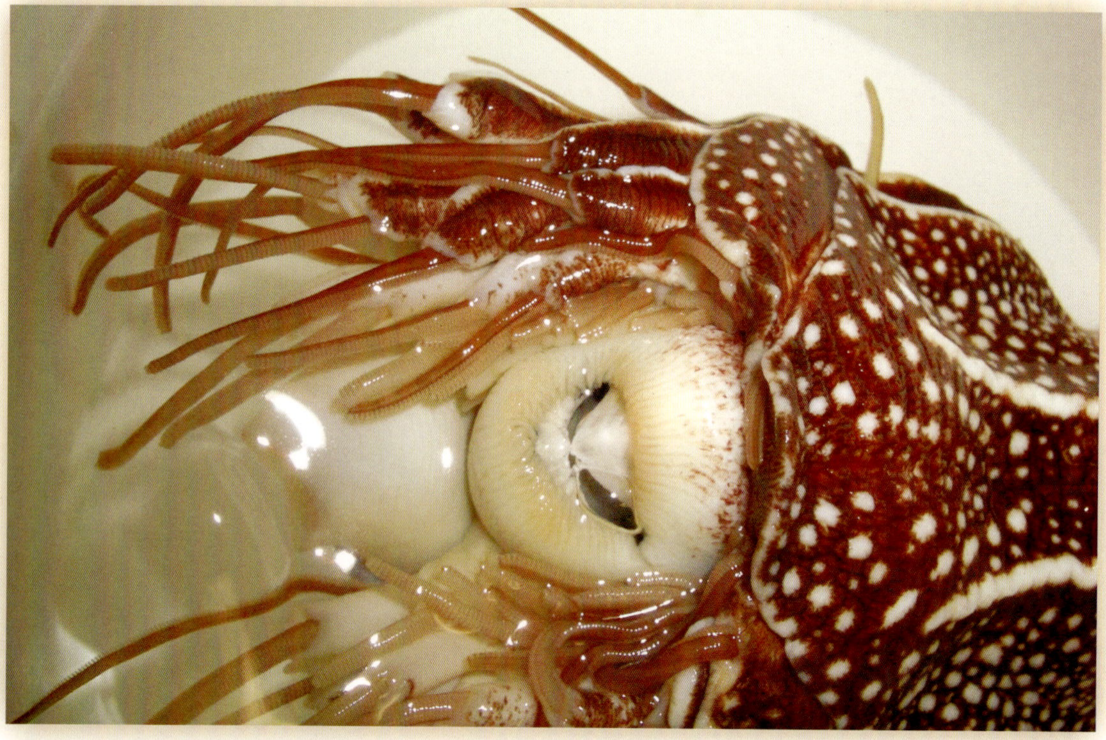

The beak and tentacles of this large female *Nautilus pompilius* from Vanautu (a small island in the South Pacific Ocean) are clearly visible. The area above the beak (to the right in this photo) is called the hood.

fill the chambers of its shell with either gases or with liquid to remain buoyant.

When it perceives danger, it can retract completely into this chamber, tentacles and all, and it relies on its hood to protect it from predators. The nautilus depends wholly on this method of defense—it can't jet away or release ink or camouflage itself as can many of the other cephalopods discussed in this book.

A nautilus has 90 or so tentacles—not arms, like an octopus, but adhesive tentacles with no suckers. Its eyes are less developed than those of other cephalopods, and they are mainly scavengers that rely on feeling rocks or the ocean floor with their many tentacles to find food.

Some local aquarium stores feature nautiluses from time to time, usually *Nautilus pompilius.* Aquarists are captivated by these creatures in their beautiful shells floating in mid-water or attached to the side of the tank. Unfortunately, we also frequently see nautiluses kept in conditions in local fish stores that guarantee their fast demise. They are put in with tropical fish and corals, not in the cooler waters they require.

A Long-Lived Cephalopod

Nautiluses have one of the longest life spans among the cephalopods. There are reports of nautiluses living as long as 15 years, far longer than any octopus or cuttlefish.

Considered exotics, they are kept like tropical fishes even though their needs are greatly different.

Although they are often seen attached to the tank's walls, nautiluses can jet around the tank in an uncontrolled way and easily run into its sides, a problem seen with squids and cuttlefishes as well. Therefore they need a sizeable tank—for a nautilus with a 6-inch (15-cm) shell, a 90-gallon (350-liter) tank at least 24 inches (60 cm) deep and 18 inches (46 cm) wide. Adult *N. pompilius* usually reach 6 to 7 inches (15 to 18 cm), with some getting closer to 9 to 10 inches (23 to 25 cm), particularly ones fished from warmer waters around the island of Vanuatu. A tank this size is large enough for one 6-inch nautilus to move around without hitting the sides; the tank would have to be even larger for more than one nautilus. You can keep several nautiluses together, but you'll have to increase the tank size and filtration to accommodate them. Always try to get a tank that is tall and wide, unlike the preferred tank for an octopus, which is longer and wider, but not particularly tall.

One advantage of keeping a nautilus is that the tank doesn't have to be a species tank, like those for the other cephalopods discussed here. They have been kept successfully with certain other animals, but you need to remember that some of those other animals could also become dinner. Also, they must be tolerant of cool water temperatures. You can have a cleanup crew and keep shrimps and snails with the nautilus, although they might occasionally eat a shrimp if they are able to catch it.

Nautiluses need large tanks, especially if you are keeping more than one specimen.

Three *N. pompilius* from the Philippines being hand fed a piece of fish.

Nautiluses need a high-calcium diet for the shell to remain strong, and they need to be fed far less than octopuses and squids: a whole shrimp or piece of bony fish once a week is sufficient. Overfeeding seems to shorten their lifespan in captivity and causes their soft tissue to out-grow the shell, which is not lethal but not quite right either. Overfeeding might also be the cause of shell aberrations if the shell expands but doesn't thicken correctly.

A nautilus's shell may suffer from degradation over time in a saltwater aquarium, resulting in a black edge to the shell. This is not a health issue, but it makes the shells less beautiful. We don't know the exact cause of this problem or how to remedy it, but it might be related to frequent crashes into the sides of too-small tanks, causing repeated small breakages of the shell.

Squids

Squids are very much present in all of our oceans and in all sizes. They have captured our imagination in the form of the pelagic squid, free-swimming in the open ocean, with the most well-known example being the giant squid, *Architeuthis dux.*

There are smaller squids found in estuaries that have been kept and even bred in research labs. These are interesting because they school and can adapt to a variety of conditions. Still, they need to be kept in very large round tanks that home aquarists find difficult to provide.

It is the smaller benthic forms of squid that interest home aquarists—the bobtail squids. Of these, the most commonly kept species with home aquarists is the Hawaiian bobtail squid, *Euprymna scolopes*. This species is found along the shores of the Hawaiian Islands.

This nocturnal squid looks different from the sleek open-ocean squids and in fact looks more closely related to cuttlefishes. Like its larger squid relatives, *E. scolopes* has eight arms and two tentacles. Unlike the sleek and fast- moving pelagic squids, the bobtails have round bodies, rounded fins, and a mantle length of only a little over 1 inch (2.5 cm).

They live in shallow water only an inch or two deep (2 to 5 cm) and bury themselves in sand or mud during the day, coming out at night to feed, still covered with sand. They are easy to spot at night, especially if you use a light.

Like many of the other bobtail squids, *E. scolopes* is bioluminescent. It has a sophisticated light organ in its mantle that houses a light-emitting bacterium, *Vibrio fischeri,* which helps disguise the small squid while hunting at night. *E. scolopes* would normally be visible as a silhouette against moonlight shining down into the water. However, the light produced by this bacterial symbiont makes the bobtail squid's silhouette invisible to predators below. The bacterium *V. fischeri* is found in sea water, and young bobtail squids have a structure to trap it and acquire a culture to build up their light organ. If they are born in an aquarium, they will never find the right bacterium for their light organ unless it is introduced.

Although they can be caught and kept for a while, these little squids can be difficult to keep and don't offer you much to see. Most hobbyists would be better off with an octopus or cuttlefish.

Euprymna scolopes photographed in Hawaii.

Chapter 9

Cephalopod Problem Solver

We felt that this chapter best serves you the reader in a question-and-answer format. Everything here addresses questions we see regularly on the forums of www.tonmo.com, the website we help moderate—routine maintenance, care, special problems and situations including when your cephalopod is at the end of its life, and more.

Should I Change the Food as My Cephalopod Grows?

As your octopus increases in size and moves towards adulthood, it will be consuming a proportionally larger amount of food. Resist the temptation to offer ever larger crabs—they should be only about one-third to one-half of the mantle size of the octopus. An octopus could kill a larger crab, but there is always some risk of injury to the octopus. It can be easier to buy large crabs, as many seafood shops will stock them for human consumption, but unless you have *Octopus cyanea, O. vulgaris,* or another very large species, larger crabs are best avoided. If no other crabs are available, you can use small 1-inch (2.5-cm) carapace crabs such as fiddlers and offer them several at a time.

Cuttlefishes grow just as quickly as octopuses, and once they get to being a few months old you can start to wean them onto nonliving foods. Initially try feeding them nonliving versions of the food they are already used to. As far as food size is concerned, stick to the rule for octopuses and don't feed them anything larger than about one-third to one-half of their mantle size. A cuttlefish over 5 inches (13 cm) will start to ignore *Mysis* shrimps and is able to accept larger shrimps and fiddler crabs.

Hermit crabs make great food items for your octopus. Cuttlefishes however will not show much interest in them.

Can I Leave Live Food In the Tank?

This technique is called free feeding. It works, but you have to be very careful to remove killed and partially eaten food as soon as you see it, because it can quickly contaminate the tank. Dead clams and mussels are especially prone to polluting the water, and it is better not to leave them in the tank for long; they tend to bury themselves in the substrate and die unnoticed.

For more information on feeding, see Chapter 6.

What Should I Do If My Octopus Has No Interest in Food?

If your octopus is still settling into a new tank, it is normal for it to not be interested in eating. It will begin eating within a day or two at most.

Are you *really* sure the octopus isn't eating? Many small octopuses will feed on the amphipods in your aquarium and may not take the food you offer them until they have grown a bit. If you have an established octopus you need to look into this a bit further. Perhaps the octopus just needs more variety in its diet. However, refusal of food can signify some form of illness, poor water quality (an ammonia spike), or even senescence (more on this topic later on in this chapter). First try to offer some other types of food, especially some live food. If your octopus eats mainly shrimps and crabs, maybe a freshly cracked mussel will tempt it. A small crayfish might be just what your octopus needs. Or it might be interested in live a shrimp, if normally fed on shellfishes. The point is to vary the diet and try something new.

At the same time, check your water quality to make sure that it's near perfect, and perform a water change if any parameters are off. Water changes are usually the best way to start making corrections in your water quality if it has deteriorated. Make sure that the new salt water is an exact match in salinity and has aged at least a day. Avoid freshly mixed salt water, as it can be harmful if added too soon to a cephalopod aquarium.

Examine your octopus; what is its behavior? Is it hiding, lethargic, or behaving abnormally? How old is your octopus? Could this be senescence, a condition that occurs at the end of an octopus's life? If so, your octopus may show physical deterioration and might act disoriented. At the very end of its life, an octopus usually comes out into the open and does not die in its den.

Are there any visible abnormalities, such as an eye infection, an abrasion, or a wound? (More on this later on in this chapter.)

Are there any husbandry issues that could be affecting your octopus? Is the temperature correct for the species? Are the lights too bright or are they on for too long? Check the photoperiod and make adjustments if necessary.

Another reason why your octopus might be off its feed could be because it is a female and has laid eggs in the aquarium. Is she hiding in just one spot, secreted away in a cave? If you look carefully you might be able to catch a glimpse of eggs. If so, refer to Chapter 7.

How Do I Clean and Maintain the Tank With My Cephalopod Still In It?

Just as with all other aquariums, you need to perform regular water changes and other maintenance to keep the tank clean for your octopus or cuttlefish while it remains in the tank. At first it's a challenge to not startle or frighten your cephalopod with the siphon tube or algae scraper, but octopuses find it interesting and try to interact with you as they become used to this procedure. They may try to capture equipment, climb on your hand, or just generally be a hindrance, although they certainly seem to enjoy it all. An octopus that one of the authors, Nancy, kept liked to come out from hiding during tank maintenance and seemed to want to participate in the cleaning. She sat in the front of the tank while the new water was being added.

Because all tank maintenance must take place with your cephalopod right there in the middle of things, be very careful to have no sharp edges such as razor blades (commonly used to scrape off algae) or anything else that could harm it. And be sure you always wash your hands and then thoroughly rinse them before putting them into the tank. Some hobbyists wear special gloves specifically sold for marine aquarium use.

During the maintenance you will normally be opening or removing the top of the tank for access while

Some octopuses will spend a great deal of time just hanging on the front glass.

doing water changes or cleaning. This is a time when your octopus can crawl out of the tank. Pay attention when the top is open, because it's easy to get distracted (a phone call, for instance) and this gives your octopus the perfect opportunity to escape. An escape can happen quickly, so make sure the tank is closed if you have to leave in the middle of things.

Be sure that any possible escape route is properly secured. Octopuses are notorious escape artists.

Where Is My Octopus? It Seems to Have Disappeared.

This is more of a problem associated with a young octopus being released within a larger tank. When octopus owners can't locate their pets for a day or two, panic sets in. The first thing that comes to mind is to dismantle the tank. This is a bad idea and will stress an octopus if it is still acclimating and settling in. There is a good chance that it is in the tank, and maybe even right in front of you. To find your octopus, you need to develop "octopus eyes" so you can scan the rocks and be able to find the hidden cephalopod. Watch the rock for the slightest indication of movement—probably the funnel just barely moving. Or you might look for its eyes. With practice, you'll spot your octopus almost every time, no matter what color and texture it displays.

It's also possible that you have a nocturnal octopus, so check at night as well as during the day. You need a red light to spot your octopus without causing it to hide or not come out at all. Try a flashlight with a red lens or red plastic over the lens.

Squirting *Sepia*

Cuttlefishes in an open tank can take aim at their keeper and be quite accurate. One of the authors of this book, Nancy, witnessed a *Sepia pharaonis* squirt her husband right in the face from 15 feet (4.6 m) away. The cuttlefishes were in a very large open tank at a breeding facility. Colin also kept *S. officinalis* that squirted him and occasionally targeted the dog.

The Times That Ollie Came to Visit

One of Nancy's octopuses, a female *Octopus bimaculoides* named Ollie, climbed out of her tank, sat on its edge, and felt all around the outside glass. She showed special interest in the cleaning magnet, which she also played with on the inside of the tank. Then she reached up and touched Nancy's husband's nose while holding tightly to his hand. Ollie made two supervised excursions out of her tank, and then she seemed content and never tried to leave again.

Rather than tearing the tank apart, it makes more sense to look all around the tank and on the room's floor to make sure your octopus hasn't escaped. Sadly, this does happen, and may happen quickly, when you have the lid open for maintenance and your back is turned. If you should find your octopus on the ground, pick it up gently and put it back into the tank—it may still be alive even if it looks dead. An octopus can live for a while out of water, depending on its size and species. Give it some time to come around even if it is looking a bit dry.

If you have a sump, refugium, or an external filter on your aquarium, give these items an examination too. It has been known for an octopus to climb over a weir or squeeze its body through an overflow pipe. One octopus owned by Colin somehow managed to escape from its tank into a centralized filtration system and then into a tank housing adult cuttlefishes, where unfortunately the young octopus met its demise. The octopus had passed through two "escapeproof" grids and squeezed its way through about 6 feet (2 m) of overflow piping, against a strong current, before getting into the second tank.

So if you've ruled out escape, just make sure there is food available in the tank and wait. Your octopus will hopefully show up after a while. Remember, most ill or dying octopuses usually come out into the open.

Why Is My Octopus or Cuttlefish Squirting Water?

Although squirting can be a form of defense, we usually see it in the home aquarium when an octopus is trying to get your attention, is excited, or is in some sense playing with you. Most octopus keepers experience this while they are cleaning the tank or just interacting with their cephalopods. Sometimes this happens when the octopus seems to be in a playful mood. There have been cases in public aquariums where a large octopus has developed a dislike for a particular person and squirts that person—and only that person—at every opportunity, but this has not been reported on the small scale of a home aquarium.

My Octopus Is Trying to Climb Out. What Should I Do?

If your octopus is at the water level and looks as if it is trying to get out, check the water parameters. Often this behavior is an indication of a problem with the water quality. Test the water; if there is a problem, carry out a partial water change immediately. Follow up until the parameters are back to normal. During this time, keep an eye on your octopus and make sure it doesn't crawl out of the tank.

A juvenile *Abdopus aculeatus* on a reef at low tide, Lizard Island (Australia). You can just barely see this octopus.

On the other hand, if your octopus comes over to investigate when you have the tank open and looks as if it wants to climb out and visit you, that may be exactly what it wants to do.

My Octopus Has Escaped! What Can I Do?

If your octopus does escape and you find it on the floor, put it back into the water immediately, even if you think it could be dead. Sometimes a presumed-dead octopus does revive. An octopus does not die immediately from being out of water. In fact, some species leave the water and travel from tidepool to tidepool, so they have the ability to be out on land for short periods. It's difficult to say exactly how long they can tolerate being out of water, perhaps half an hour for the species commonly kept by home aquarists.

If you can't find your octopus, look all around the tank, then search a bit further. There are reports of a dwarf octopus that made it all the way from its tank to the front door of the house, then under the door and down the driveway to the mailbox on a hot summer's day before succumbing.

Why Won't My Octopus Come Out of Its Den?

There are a number of possible reasons for this. First of all, could you have a nocturnal octopus and you're trying to view it during the day when it stays in its den? Check again at night using a red light to see whether it's active.

Is your octopus newly arrived and not used to its new surroundings? It may take a few weeks for it to feel at home and comfortable enough to come out and explore more often. An octopus needs to feel secure in its new surroundings.

Are you using metal halides or other bright lights? If the lights are too bright, an octopus will stay in its den, not venturing out all that much. Switch to a regular fluorescent bulb or even ambient light for a few days and see whether the behavior changes.

Do you have fishes in the tank? Fishes can make an octopus stay in its den, especially so with younger octopuses. Damselfishes in particular are noted for going after an octopus's eyes. We mentioned keeping your octopus by itself in a species tank, and this is one of the reasons why.

Could your octopus be a female that has created a den for laying her eggs? Is your octopus an adult of the species? Ideally you need to know what species you have to determine this or at least whether your octopus is no longer a juvenile. Egg laying usually occurs towards the end of their lives. After laying the eggs and taking care of them until they hatch, the female octopus will not live much longer (at least in most cases). There are a very few species that will lay successive batches of eggs, but they are exceptions.

What Should I Do When My Octopus or Cuttlefish Inks?

All octopuses that we keep in home aquariums have a brown or black ink that they can release for defensive purposes. If they get stressed or afraid, they may ink. Some species of octopus, and certain individuals within those species, ink more than others. Many will ink during acclimation, especially when they slide into the tank for the first time. It would be normal for your octopus to ink several times during its life with you.

The ink is not deadly to the octopus, although it should be removed as quickly as possible. This is one reason why we recommend a good-quality over-specification protein skimmer for every octopus tank. The skimmer will do an excellent job of pulling the ink out of the water. In addition to the skimmer, you can replace carbon in the filter, because carbon will also remove ink through adsorption. A partial water change is also helpful.

You can experience inking in the home aquarium in two forms. The first is ink with a thick consistency, so much so that it can be removed with a fine net, or a net with pantyhose material fitted into the insides. The second form is a smokescreen of ink that must be removed by the skimmer and carbon because it is too thin and widely dispersed to be caught in any net. Although ink is almost pure melanin, it can contain mucus—this is what makes the ink stick together in a thicker form. If an octopus inks in a confined space such as a transit bag it might die through asphyxiation.

Cuttlefishes, especially *S. officinalis,* can literally black out a tank with their ink. A 6-inch (15-cm) individual can produce enough ink to make it almost impossible to see into a 200-gallon (800-liter) aquarium! In these cases it might take a protein skimmer and carbon at least 24 hours to remove the ink.

Why Is My Octopus Missing Arms?

Sometimes you may buy an octopus that is missing an arm or two and you might not even know this until you have the octopus home. This disfigurement may have happened during its capture; more likely it was collected this way after an incident in the wild. Losing an arm or two is a defense strategy of octopuses; it allows the octopus itself to escape a predator. Some species, such as *Ameloctopus,* are more prone than others to do this. If the octopus is otherwise healthy, the arms will grow back, although they may always be slightly shorter than the original arms.

It sometimes also happens that an octopus that came with all eight arms now has one of them missing—it may be the whole arm or just part of the arm. This could be because of an encounter with a crab, perhaps one that's too large for your octopus to subdue. It could also be from sticking an arm into the filter or powerhead in your tank while exploring and looking for food. Check to make sure all intakes and

Seeing their octopus curl into a ball can cause a new octopus keeper unnecessary distress—this is how many octopuses shed their sucker discs.

outlets are protected. Again, the arm will grow back, but just make sure that arm losses don't continue by eliminating the cause.

There is also an uncommon condition called autophagy, in which an octopus eats its own arms. Sometimes this seems to be caused by stress, and the octopus will recover. Sometimes it has a natural cause, possibly a prion, and is highly infectious. If your octopus has this latter form of the disease, the tank will have to be sterilized after its death.

What About Other Injuries?

An octopus can be injured in various ways. One of the most common is to be pinched by a crab or perhaps to obtain an abrasive injury from a rock or something sharp in the tank. If, contrary to recommendations, you have fishes in the tank, your octopus could be bitten. Sometimes octopuses get skin lesions from rubbing against live rock. If the octopus is otherwise healthy, it should recover from these injuries. Make sure your water parameters are excellent to aid in recovery.

Cuttlefishes can easily suffer abrasions from hitting or rubbing against the aquarium glass or rocks. Although such damages could possibly allow infections to set in, most cases seem to heal on their own. Cuttlefishes can also rub away the skin on their tentacles, perhaps reacting to seeing their reflection in the glass. Most recover from this, and it doesn't seem to bother them much.

Your octopus will eventually come out into the open and observe all the strange creatures on the other side of the glass.

A displaying female *A. aculeatus.* The small raised pieces of skin, called papillae, are typical for this species.

How Can I Treat an Octopus or Cuttlefish With Eye Problems or Other Infections?

Cephalopods often have eye problems, especially as they grow older, when an eye can take on a cloudy appearance. There are reports that certain remedies are effective, but you must determine after careful study whether it's appropriate to give a cephalopod antibiotics and also to see how this can best be done. If the cephalopod is approaching senescence, it will not do much good to provide treatment. The best place to research possible medications is online. In some countries antibiotics must be prescribed by a veterinarian, but in the U.S. some aquarium-grade medicines can be purchased in pet stores. If you decide to take this route, be sure you administer the correct dosage. Most likely you'll have to treat the whole tank or administer the medication into food.

When treating sick cephalopods, it is prudent to do regular partial water changes and make sure the water is pristine. This will help the immune system combat the illness.

Octopuses that become blind can still function because they receive much of their sensory data from their arms.

How Can I Prepare My Cephalopod Tank For When I'll Be Away?

Any saltwater tank could be subject to a catastrophe if you leave it unattended for longer than a couple of days. You can provide some extra food in the form of a few crabs to make sure your cephalopod can get along for a couple of days, but leaving it for a longer period of time can be risky. For instance, you might have an equipment failure, such as a water pump jamming stuck after a power failure, or a heater malfunction. Or perhaps the octopus will ink in the aquarium or leave uneaten an item of food that will quickly rot, and your water quality will deteriorate.

No matter what types of home aquaria you have, it is notoriously difficult to rely on a friend or relative to look after aquatic animals. They mostly have no animal husbandry experience whatsoever—they will misunderstand, overfeed, underfeed, and not know a thing about maintaining water quality.

One of the best solutions reported by cephalopod owners is to have one trusted person come in every day to feed the octopus (or other cephalopod). Have an experienced reefkeeper also come in every day to make sure the whole system is working right and that water quality is good.

Be aware of the possible effects of plugging your equipment into a ground fault circuit interrupter (GF[C]I) that does not automatically reset, in case you have a power failure. These circuits are good protection for you and your tank, but leaving a tank connected to a GF(C)I that does not automatically reset means that your equipment will be off until someone manually restarts it. This is the best safety option, but many a tank's livestock has been lost due to power failures under these conditions. You will need someone to check the tank daily to make sure this hasn't happened if you are away.

A second concern is equipment, especially pumps, not restarting when the power comes back on. This happens more in the case of older pumps, which may have worn impellers and not kick into action when the power returns. They can usually be restarted by cleaning them regularly and paying particular attention to moving parts and ensuring that sand or grit doesn't get stuck in the impeller housing. Shaking them or knocking them against your hand might work, but if no one is there to get the pump started again, the tank is left without filtration. If you begin to have problems restarting your pump, it's time to think of getting a new one.

How Do I Handle a Power Outage?

You will most probably encounter power failures during the time you keep a cephalopod. You need to plan for this eventuality and know what steps to take, whether it is summer or winter, to keep your cephalopod safe and healthy.

In the case of a power failure, your main concerns will be oxygen content, keeping the bacteria in the filter alive, and the water's temperature (whether it is heating or cooling). Most power failures are for a limited time, and most likely the power will be back on before filtration becomes a problem. In one case of a repair to the power system, an octopus owner went over nine hours without power, and her *O. bimaculoides* survived quite well. Don't worry about the tank lights, because an octopus does not require them.

There are auxiliary power supplies that you can buy—some for your entire house, others more limited in scope. Some are generators, others more like battery backups.With this auxiliary power, you can operate your filters and pumps as normal. Battery-operated air pumps can provide auxiliary aeration during an outage.

If you don't have this extra equipment, there are still some things you can do. Assuming that you can do without filtration for a little while, your immediate problem will be tank temperature. If you have *O. bimaculoides* or any other cephalopod from cooler waters, it will tolerate lower temperatures (at least the upper 50s), so it's the summer power failure that's a problem for that species in areas that have hot summers. For tropical octopuses, heat needs to be retained in the tank. You can insulate the tank with newspaper, cardboard, or polystyrene by using duct tape to hold it around the tank. This will slow down further heat loss.

You can add heat by using plastic bottles or jugs full of hot water. Most people have some form of auxiliary heat, such as a gas grill or a propane stove or even an open fireplace, so that the water for the bottles can be heated. Float the heated bottles in the tank. It's best to use RO/DI water in the bottles.

In case of a summer power failure, you can fill the bottles with ice water from your refrigerator. Keep some ice cubes on hand to use for making cold water to fill bottles. Insulation works for retaining cold as well as heat.

To add oxygen to the tank, you can fill a gallon jug halfway with tank water and then shake it vigorously

to aerate it and pour it into the tank. Oxygen sources like a tank of oxygen can be used, but you'll need a regulator.

All this shouldn't be necessary for a short power failure, but it's useful to know in case of a major problem.

What Does It Mean When My Octopus Has Curled Arms?

Octopuses will show various types of behavior, but if you see something such as an older octopus sitting out in the open with curled or "corkscrew" arms, it could be a sign that your octopus is near death. However, some species curl their arms more than others, so it can be perfectly normal, even in a younger octopus. An octopus rubbing its arms to rid itself of old sucker disks may also have curled arms, but the difference is that it stops this behavior and goes about its business. If you have noticed other signs of senescence, this is almost surely the case. The end comes quickly, and unfortunately all you can do is wait.

In other cases, twitching arms and an apparent restlessness can also be a sign that something is seriously wrong with water quality. An octopus showing signs of heavy metal poisoning can quickly succumb if added to an aquarium affected by the contaminant. Be aware that cephalopods are very sensitive to copper and other metals and show their discomfort with arms that seem to be in charge of themselves.

Is My Octopus or Cuttlefish Reaching Senescence?

The octopuses and cuttlefishes we commonly keep in our home aquariums do not have long life spans. Depending on the species, they may have a total natural life span from six months to a little over a year or a year and a half. In our home aquariums the cephalopod may already be a few weeks or months old by the time you acquire it, therefore shortening an already brief life expectancy.

Senescence is a stage at the end of your octopus's life, often occurring a couple of weeks or even just days before it dies. Signs of senescence will vary according to species, sex, and even the individual. Much about senescence has not yet been studied, but you'll have little difficulty recognizing the symptoms when the time comes.

Although both males and females enter a senescent stage before dying, it is different for each. Males may enter senescence after mating and are more likely to be active and out around the tank. Females display signs of senescence after laying eggs or at least after the eggs hatch, and are more often reclusive. Both may eat seldom or not at all, show signs in the skin (lack of color change so that sometimes they are almost white, lesions that don't heal, etc.), and suffer from eye problems because their immune systems are not as strong as they once were. In short, you'll notice physical deterioration. Movement may be more uncoordinated, and the octopus may wander about the tank. Both males and females exhibit these signs.

It's important to understand that senescence is a normal occurrence at the end of an octopus's life and not a disease, although disease may also come into play because of the weakened immune system. The only thing that helps is to keep offering food through the female's brooding period (even if it has to be delivered to her den door using a feeding stick) and after the eggs have hatched, and during the onset of senescence for males as well. Know, though, that offering food will not prevent senescence.

The most bittersweet aspect of the cephalopod hobby is that most octopuses and cuttlefishes live only for a year or so. This octopus sadly has died. During senescence and death an octopus usually loses all colors.

Sometimes near the end the octopus's arms assume a corkscrew-like form, more a sign of the end than senescence (see previous question).

This is a sad time for a cephalopod keeper, and there is little you can do to help. In the wild the cephalopod would quickly be eaten by a predator when it is in this state.

My Octopus Has Died.

When the gloomy day comes that your cephalopod dies, you'll first need to determine the cause of death, if you can. What were the water parameters at the time of death? Could this be the natural end of life after a period of senescence, or did it come early and unexpected because of some external influence or internal condition? You may never be able to determine the exact cause, but you should always try.

It is best to clean the tank after a cephalopod's death and perform a major water change. You can get another cephalopod within a short period of time unless there was an infection or autophagy. Chances are that the water quality is not as perfect as it was when you first got the octopus, so be patient and wait until parameters are correct again.

A large dead cephalopod, especially a cuttlefish, can expel a large amount of ink as it dies and make a real mess of its tank, to the point of making it impossible to see into it.

Chapter 10

Photographing Your Cephalopod

If you keep an octopus or cuttlefish, you are bound to want to photograph your cephalopod doing all the interesting things cephalopods do. Technically you can just point and click, but there are other things to take into consideration to get the best photograph possible.

THE following suggestions have been put together after reviewing hundreds of photos of cephalopods in home aquariums that have been posted on Internet forums. We can tell from these photos that cephalopod photography takes some thought and planning—and that cephalopods are not easy to photograph! The information provided here will help you understand how to approach your subject and offers tips for better cephalopod photos. While it is not by any means comprehensive, it is a good starting point for learning the bare-bone basics of cephalopod photography.

Taking photos of your pet octopus or cuttlefish ranks high on the list of favorite things for cephalopod owners to do. But it's not always easy to do. Cephalopods can be difficult and reluctant subjects, and their camouflage abilities make a good photograph even more complicated to achieve.

They need to be photographed at close range, which leads to problems with focus and composition. The most challenging subject is the octopus with its many arms extending in all directions, making focus and depth of field difficult. Tank lighting and nocturnal cephalopods make taking good photos even more complicated.

Many of the pictures we most admire are taken by professional photographers or researchers with high-end equipment. Some are pictures taken by divers with underwater cameras utilizing special lighting, while others are taken in the lab during long photo sessions using strobe lighting. Nevertheless, excellent photographs can be taken by octopus and cuttlefish owners with any of today's modern digital cameras. In fact, even a camera phone can occasionally show off a cephalopod well.

Most cephalopod keepers opt for a good point-and-shoot camera with enough control to take tank photos, but it's certainly still possible to take interesting and very detailed photos with a 35mm camera using slide film. In this modern digital age this is becoming somewhat rare—and markedly more expensive.

LEARNING FROM OTHER CEPHALOPOD PHOTOGRAPHS

Over the years we have looked at hundreds of cephalopod photos posted on the Internet by their owners, and we've noticed several recurring problems. The most common problem was in the focus of the picture.

Although some photos were entirely out of focus, far more common was for part of the photo to be in focus and the rest blurry. The part of the photo in focus was often not the center of interest, so the effect was to emphasize a minor element in the composition. A second problem was composition, and this occurs more with octopuses than with other cephalopods. To be recognizable, an octopus must look like an octopus, not just a blob or tangle of arms. The third problem was unnatural colors, usually caused by lighting issues.

Nautilus, Argonaut, and Squid Photography

Nautiluses don't present many of the same problems that octopuses and cuttlefishes do and are more straightforward to photograph, although they are often photographed in groups. Since squids and argonauts are rarely kept in home aquariums, they won't be covered in the following recommendations.

In this picture, cephalopod enthusiast Richard Ross is photographing a flamboyant cuttlefish. He is using a hand-held external flash with his camera. Cephalopods don't generally stay all that still, so having a flash that moves can be very helpful. He is also on a stepladder to get properly aligned with the tank glass to avoid distortion and he has all the blinds in the room closed to avoid reflections.

YOUR CAMERA

Whether you already have a camera or are buying a new one, here are some features to look for that will help you take good photographs of your pet cephalopod.

Most pet owners use digital cameras with autofocus and fully automatic settings. However, if they can afford it and can tolerate the learning curve, they find that digital single-lens reflex (SLR) cameras are a better option. In any case, the more megapixels (MP) the camera has, the better the detail in your photos.

Camera Design

Getting a good grip on your camera may be difficult with some of the very small cameras. There will be times when you need to grab the camera and start shooting in order not to miss a spectacular shot of your favorite cephalopod in an instant. If the camera is too large and heavy similar problems may occur, so make sure you can hold your camera comfortably. An automatic stability feature (vibration reduction), available in either the camera or in the lens, depending on the manufacturer, will assist with quick shots where you don't have time for a tripod.

Good-Quality Lenses

This is an important aspect in regard to getting sharp photos. Good-quality lenses are standard in reasonably priced compact cameras. If you can, choose optical zoom over digital zoom for better photos.

If you have a digital SLR camera, you have the luxury of being able to change the lens to suit different circumstances. With this in mind you can have a wide angle for full tank shots and a dedicated macro lens for close-ups; then you can get much closer to the action and see the really small details of a cephalopod's skin. However, some photographers get excellent shots with an 18-55mm lens, which often comes standard with the camera's body.

Some Focus Control, Even With an Autofocus Camera

The ability to manually control the camera's focus will be important in shooting through glass. You need to be able to focus on the cephalopod, not the glass or any algae in the tank.

You'll need to master the control over focusing that your camera provides and understand the options it offers. Many professionals feel you will get better photos if you use manual focus and don't use automatic focus, but that takes more control and a lot of practice on your part. You also need to learn about depth of field, which is the area in which your subject will be in focus.

Although not recommended for the home acquarium, you can clearly see why the wunderpus was given the name *photogenicus*.

In this photo, the camera focused on the rear wall and the flash has bounced off the back glass. With a little practice you are sure to capture a crisp, stunning shot of your cephalopod.

Judging from the cephalopod photos contributed to websites, focus is the greatest problem cephalopod keepers have with their photographs. This is more noticeable in photographing octopuses than cuttlefishes or nautiluses. The front of the octopus looks fine, but three or four arms will be out of focus. Sometimes this can be intentional when zooming in on a particular feature of your octopus, but most photographers will want the entire octopus to be in focus, so this is worth further practice and study.

On a digital SLR you can alter how much of the octopus is in focus by adjusting the f value. The lower the f-number, the larger the camera's aperture, but less of the animal will be in focus. To get the whole animal in focus you will need an aperture value of perhaps f/16, but you will then probably have to use a flash, since the shutter speed will be slower. Ideally you should shoot in fully manual settings. This will take a lot of trial and error and a bit of research!

Usually by taking a number of quick photos in succession, and by moving your position slightly at the same time as the cephalopod moves, you can get at least a few decent shots in good focus. Professional photographers take large numbers of shots and choose the best, so don't feel bad about discarding most of them. That's the beauty of digital photography: you can take tons of shots until you get the one you want without spending a small fortune on developing film!

With Flash

Sometimes it is necessary to lighten up the photograph by using flash photography. There are a few complications with using a flash in front of an aquarium, such as light bouncing back off the glass to obscure the tank in the photograph. Worse still, cephalopods often react adversely to the flash by becoming frightened and inking in the tank. You must use the flash feature carefully and don't do anything that will stress your octopus or cuttlefish.

No-Flash Option

Some excellent cephalopod photos have been taken without using the camera's built-in flash. The flash feature should be easy to de-activate, perhaps by pressing a button rather than by working through menus.

In order to get good photographs in low-light conditions you normally have to use a long exposure. This, however, will often give you blurry pictures because the octopus will move about. One of the few ways to get clear, sharp pictures without the flash is to make sure that the tank is lit by good overhead aquarium lights.

White Balance

The ability to directly adjust white balance is becoming common even on less expensive point-and-shoot

You've captured a shot of your beautiful bimac, but your tank's glass caught the flash. Try again!

K Values

- Incandescent: Light 3000 K
- Fluorescent (cooler color): 4000 to 5000 K
- Electronic Flash: 5000 K
- Daylight, clear sky: 5000 to 6000 K
- Overcast: 6000 to 8000 K
- Shade or cloud (warmer color): 10,000 K

cameras. White balance settings are used to remove unnatural color casts that can produce unnatural-looking photos. White balance takes into account the color temperature as well as the green-magenta shift.

You can recognize tank photos in need of white balance adjustment—sometimes the whole photo is too red. (This of course excludes photos taken of nocturnal cephalopods using a red light.) A white balance setting enables you to adjust for different types of light, including the lamps you're using in your tank. You may have to experiment with the setting to find what works best for your situation. Normally you can choose between sunlight, shade, tungsten, and fluorescent settings, all of which will give you different results.

Some cameras have a customizable white balance setting that can be set by focusing on a white background with the lighting in question. Others will just give you some options. If you have the customizable setting, use a white surface to focus on and then carry out the customization process according to your camera's instructions. Another option is to buy an external white balance filter, which is held up in front of your lens.

It's useful to know the temperature rating of various light sources, because then you'll know the approximate K (Kelvin) value of the lamp you're using on your cephalopod tank.

Here are common K values that correspond to white balance settings on your camera. These figures are general; different lists vary somewhat. In other words, if you have a 10,000 K lamp on your cephalopod tank, which is a color temperature sometimes used for cephalopods, try the shade or cloudy white balance setting on your camera. If the colors are too cool (blue), try a warmer setting (higher K value). This gives you at least some control over white balance.

Try the custom white balance setting if you have problems achieving an accurate color temperature. Note that people with reef aquariums using actinic or metal halide lamps have more trouble with color balance than people with the simple lighting normally used in a cephalopod tank.

If your camera can take photos using the RAW setting you can adjust the white balance with software that

An excellent shot of an *Octopus bimaculoides*. Look at the beautiful colors around the suckers.

comes with the camera or bought separately after the photo has been taken. Many cameras will come with their own software to install on your home computer to help you edit pictures once you download them from the camera or the camera's memory card. Ideally you should shoot in a mode that takes a JPEG and a RAW picture at the same time. With this setting you can see the JPEG at a glance and later tweak the pictures you like in their RAW format, which gives you much more control over the end product because you can change many features. RAW format means that the camera makes no decisions by itself and the tweaks are entirely up to you.

Low-Light Ability

You can adjust your camera to work in low light in various ways—choosing a higher ISO (film speed) setting, for instance. Some cameras have specific low-light settings that you can easily select, or you can adjust the photo after it has been taken using software.

Colin's Photography Settings

One of the authors of this book, Colin, usually shoots at 125 to 180th of a second at f/11 to f/16 with a flash positioned above the aquarium on a sheet of plastic grid triggered by a wireless remote on the top of the digital SLR. He always shoots in RAW format and mostly with JPEG at the same time. Colin has found that most of his pictures are taken at around a 50-mm focal length but that a 105-mm macro lens is ideal for closeup shots.

When photographing in a lower light, you'll need a larger lens opening (lower f-stop number), which shortens the depth of field and can result in focus problems. The slower the shutter speed the more light will reach the sensor, but with this there's a higher chance of blur. The quicker the shutter speed the more light is needed, and there's less chance of blur. The lower the f-stop number the more light gets into the camera, therefore the faster the potential shutter speed and less chance of blur. It all sounds confusing, but trial and error as well as a good instruction manual will eventually explain what is needed for aquarium photography.

Even if you overcome many of the problems of low-light photography, the available light may not be adequate for taking good photos of nocturnal cephalopods without a flash. Consider using a remote flash from the top or side of the aquarium, and a light diffuser of some sort. You can also use a tripod with a slower shutter speed if your subject will remain still. Some of the more recent digital SLRs can use a high ISO (800 or more) and enable you to take excellent tank shots without a flash.

LEARNING TO USE YOUR CAMERA'S FEATURES

Of course, you need to be familiar with all the features of your camera and their use. Failure to master your camera and its features properly is usually the cause of less-than-satisfactory photos. Cameras are complex today, even the less expensive ones. To make good use of your camera's features, you need to

This photo captured a bimac in motion.

understand them and know how to activate (and deactivate) them.

You can miss good shots if you have to refer to the manual too much or you have difficulty understanding the manual. Be sure to read it thoroughly until you understand all of your camera's features. It may help to buy an additional book about your camera and about photography in general. Other helpful resources are the photography forums on aquarium websites. Many of the problems and solutions discussed on these forums will apply to taking photos of your cephalopods.

USEFUL ACCESSORIES

There are certain accessories you can buy for your camera that will enable you to take better photographs of your cephalopods.

Hood

This fits on the lens and enables you to get very close to the tank's glass while eliminating light and reflection. The hood can be pressed directly against the glass. A lens hood can help you get reflection-free photos, assuming that the lens has the correct focal length.

Tripod

A tripod is not as necessary if your camera has a stability feature, but it will enable you to focus and get the photo you want without worrying about holding the camera still during longer exposures.

For planned photo sessions, consider using a tripod for stability and to give you time to plan and compose the shot. It also enables you to take a number of photos of your cephalopod from the same point. You might even keep the camera and tripod set up near your tank.

The tripod is very helpful for holding a remote flash, which can be useful when doing close macro shots.

Remote Flash

Wireless remote flashes give you more flexibility by adding a flash that can point in another direction from the camera and allow you to move around the aquarium without getting tangled up in wires. Aquarium photographers often place a remote flash above the tank, shooting downwards to maximize the light above the subject or place it to the side of the tank, all of which helps reduce bounce-back of the flash on the glass.

Photo Tip

To absolutely avoid reflections from the room or outside, shoot in the dark with only the tank lights on and possibly a remote flash.

This cuttlefish has flattened itself into the tank's substrate. Interesting natural behavior equals perfect photographic opportunity!

TANK CONSIDERATIONS

If tank photography is important to you, take that into consideration when initially purchasing and setting up your tank. If you are reading this book and already have a tank with a cephalopod, this section still provides some suggestions for creating better photographs.

Choosing a Tank That's Good For Photography

If taking good cephalopod photographs is a priority for you, avoid bow fronts or any tank with curved glass or round corners. Rounded glass will distort the inhabitants and make them look unnatural. It's not impossible to take good photos of cephalopods in these tanks, but it's harder to do so.

The glass in certain tanks is optically better than in others and worth researching. Optically clear glass is increasingly available, especially for the front of the tank. Acrylic tanks can be easily scratched, and these dings will always show up on photographs. Professional or scientific photos of cephalopods are sometimes taken in special tanks with thin glass to make the end result clearer. Before taking any pictures make sure that the glass is spotless, both inside and out. You can use a new razor blade to clean the inside and a lint-free cloth on the outside of the tank.

Placement In the Room

The placement of your tank in a room in your home with respect to ambient light (especially with sunlight coming through windows) will influence the quality of your photographs. At least you can turn off room lights, but it's difficult to block light coming in from a large window. Also, the glass in your tank may pick up reflections, which are very distracting in the photos. You may even see your own reflection in some photos. Take the placement of the tank in the room into consideration when you plan your tank. If reflections are a problem, try closing the blinds and curtains, taping a dark-colored sheet over the window, and turning off all room lighting. You can also use a screen made of foam board or other material to block window light.

If you are taking pictures using a camera with a built-in flash, you will get a flash reflection if you shoot straight onto the tank. To help avoid this you can try taking the picture at a slight angle, although don't overdo it—you can go too far and distort the shape of the subject.

Clean Glass

The glass must be scrupulously clean to get the best results. A razor blade scraper will do the best job, better than a cleaning magnet. If you're going to take photos frequently, keep the glass clean at all times. Don't forget to clean the front of the glass as well, using a mild cleaner (spray this on the cloth, not on the glass). Make sure that you don't accidentally drop the razor blade into the tank and lose it, as the rusting metal is toxic to cephalopods.

Clean Water

In a typical cephalopod aquarium there is a lot of activity from the animals, and this can keep small particles of detritus in suspension. If taking pictures, especially if using an overhead light, you will be able to see all these bits of dirt and sucker cups floating about in the water. Therefore the fewer the small particles in the water, the clearer the photo will be. Polishing the tank water prior to your photo session (starting the night before) will give you the water clarity you need. Use extra carbon and a very fine filter medium to achieve this state.

Backgrounds

If all your tank's walls are made of clear glass, so much so that you see the room's walls through the back of the tank, you may want to use a sheet of colored paper (perhaps blue or black) or another backdrop behind the tank to give it an uncluttered, aesthetically pleasing background. Situating the tank in front of a colored wall can produce an interesting effect as well.

On the other hand, if you already have an aquascaped tank and a painted or covered back wall, there are still a few things you can do to enhance the environment for photos. You can add colored items, such as shells or marbles—this will add color interest to the tank for your photographs. Also, colorful toys may grab your cephalopod's (especially your octopus's) attention.

The back wall of the tank will also reflect the flash back to you, so it's best to shoot at an angle other than a right angle to the back tank wall.

TAKING BETTER CEPHALOPOD PHOTOS

You can take better photographs in one of two ways: by having your camera on hand and shooting a lot of photos of your cephalopod at various times of the day, or by trying to set the stage for a good photo shoot.

Feeding time guarantees a lot of activity, so this is a good time to have the camera nearby. Some cephalopod owners always have the camera handy when feeding, even when they don't intend to take photos. That way their octopus or cuttlefish becomes used to the sight of the camera and won't be frightened by it. Some very good photos have been taken using this approach.

You may need someone to help you with your photo session, someone to take the photos while you interact with your octopus or cuttlefish. Take a lot of photos in each session and choose the best.

What Kind of Photos Should You Take?

Certainly some pictures will show your cephalopod at rest, an octopus on the glass or in his den. However, the most interesting photos will be action photos, in which your cephalopod is actually doing something: swimming, catching prey, taking food from a feeding stick, exploring over the rocks, holding a toy or the cleaning magnet. If you keep several cephalopods in a tank, such as small cuttlefishes, their interaction can make an attractive photograph.

This wunderpus's tank is located in front of a dark red wall, producing a striking background that emphasizes the colors of the octopus.

Some of these photos can be staged, such as taking many during feeding time or giving your octopus a jar to open, but others you can catch by being in front of the tank with your camera at the right time. This is why it's a good idea to keep your camera ready and near the tank at all times so you won't miss out on any good shots.

Experiment

You'll have to experiment to find the best settings for your camera and your tank. Once you master focus through the glass, you can try flash or no flash, different white balance settings, and various ways to stimulate your octopus (toys, food, etc.). Digital cameras make all this experimentation easy because you have instant results that you can review on the preview screen and delete if they are unsuccessful.

Experimenting with cameras can also prove valuable. One octopus keeper began using an underwater housing and submerging his camera in the tank. The results were excellent, even better than shooting through glass. Fortunately, he had acclimated his octopus to the camera, so there were no issues with the camera's being in the tank.

Make Sure Your Cephalopod is Used to the Camera

Some of the best photographs of cephalopods in the home aquarium have been taken by hobbyists who

A brilliant macro shot of the suckers of an octopus.

This remarkable photo of an *Abdopus aculeatus* by animal behaviorist Roy Caldwell was edited to remove a bit of background "noise." He puts octopuses in a special photography aquarium with a flat black background inside the tank. He shoots with twin strobes and a medium length macro lens (105mm). A piece of black cardboard fitted over the lens reduces reflections. Roy feels the keys to outstanding octopus close-ups are clean water, clean glass, proper lighting, and patience.

make sure their cephalopod is not afraid of the camera. How do they do this? Simply by bringing the camera out whenever they're around their cephalopods. Even when the cephalopod is young it will get used to seeing the camera. Otherwise, the mere sight of a camera may arouse fear, because it is something new and strange. This is especially true of cameras that use a light for focusing on objects in low-light conditions—the sudden flashing light can cause alarm.

Take Photos Over the Lifetime of Your Cephalopod

Keeping a cephalopod is a special experience, one that is worth recording. A mistake many cephalopod owners make is to take initial photos only at the time their small cephalopod arrives. If you have an octopus, this can be difficult, as it may go into hiding for a while. Then a long period of time goes by, and finally there's another set of photos. You'll be happier with the results if you make documenting your cephalopod's life a continual process so that you have photos of the arrival, its early months, playing, eating, exploring the tank, and so forth. In fact, you can use the photos to accompany an account of your cephalopod's life. Such an account is welcome on cephalopod websites and makes a wonderful photo-book for you.

Take Photos in Public Aquariums

You may also have an opportunity to take photos in a public aquarium. Check first to make sure photography is allowed. Some aquariums allow photography but prohibit using a flash. It's possible to get very good photos of cephalopods in public aquariums, despite the thicker glass of their tanks. You just have to be there when the octopus or cuttlefish is in the right location, or, in the case of octopuses, visible at all.

AFTER YOU'VE TAKEN THE PHOTOS

It's best to analyze and make decisions about your photographs soon after taking them. You can learn more this way, and selecting the best, cropping, adjusting, and then organizing them gives you the best access to your photos in the future.

Improving Your Photos Through Software

You can use computer software to crop your photos and improve the contrast and color balance, especially if you shot in RAW format. There are many competing programs, so you'll have to do some research to see which one suits you the best. A word of caution: be sure you keep copies of your uncorrected original photos. If your only copy of a photo is corrupted during editing, it may be lost forever. Some advanced photography software does this for you automatically.

Also, if you ever want to publish the photo, you will need an unaltered copy. You may also need the original large-size file for making larger prints. Many magazines prefer the image to be submitted in RAW format with no editing by you at all. Again, you should choose the option that takes a JPEG and RAW file at the same time. That way you can quickly see which pictures you like by looking at the JPEG and make more complicated fine-tuning adjustments with the RAW version.

Organizing Your Photographs

You can either use your photo software for organizing your photos or set up folders on your PC. For hobbyists keeping a number of cephalopods over time, it helps to keep all the photos of an individual cephalopod together in one folder. Even though it seems like extra work, renaming your photos or adding meaningful notes will help you manage your photos in the future. You might consider renaming the photos to always include the cephalopod's name and the date.

Publishing Your Photos on the Internet

There are a number of photo-hosting websites on which you can upload your cephalopod photos. Some may have size limits on each photo, so be sure to check the site's requirements and the file type. You can also post to a photo site and link to the cephalopod site, but that will leave you with a link that goes nowhere if your photos age out and become dead links. Choose only your best photos for posting out of the many that you've taken. It also helps to identify the species (if known) and provide more information about your cephalopod.

Many cephalopod owners prefer to set up their own websites so that they can feature their cephalopods with a number of photos and videos. A few even set up webcams to record videos of their cephalopods.

Closeups of your pet can be unusual and interesting.

MAKING BETTER CEPHALOPOD VIDEOS

There are many cephalopod videos already posted on the web. Most have some problem with focus, but the main difficulty encountered is keeping the subject in the frame. Cephalopods are usually the subject of videos when they're doing something or heading somewhere. In an attempt to follow this motion, the camera often wanders off focus. This could be edited out, but most people don't bother with editing and post their videos just as they were taken. The videos' quality could be improved with practice. Cephalopod owners want to show their friends and fellow forum members what their cephalopod is doing, so these are informal, spontaneous videos and quite effective. Not to mention fun!

Appendix A

Visiting Cephalopods

If you're interested in keeping cephalopods, you might also be interested in seeing more of them and looking at species you can't keep at home. Fortunately, most public aquariums have at least one cephalopod exhibit, often of larger and more exotic species than the ones you are likely to have in your aquarium. The most popular cephalopod on display in public aquariums all over the world is the giant Pacific octopus, *Enteroctopus dofleini*. Many public aquariums could also have some other species of octopus or either a nautilus or cuttlefish tank.

THE downside of octopus exhibits is the same as you might encounter in your home aquarium: the octopuses' ability to virtually disappear, leaving you standing in front of an apparently empty tank.

By "public aquarium" we mean the privately owned aquariums that are open to the public as well as those funded by cities or other governments. Do consider the smaller aquariums, especially those along the coasts. Call ahead to find out not only the times and what's on display, but also ask about feeding and other special events, such as giving the octopus toys for enrichment. Sometimes you can also volunteer at your local aquarium and have a closer look at the cephalopods.

Another way to see a cephalopod is by walking along the shore where it might be found. For instance, along the southern California coast you might see an *Octopus bimaculoides* while admiring the sea life in tidepools. Further south, along the Baja California Pacific coast in Mexico, little *O. digueti* live in shells along the shore. Along the East Coast and Gulf Coast of the U.S. you could see the squid *Lolliguncula brevis* in estuaries. Unfortunately, there are no cuttlefishes along the American coasts.

Octopuses are not found near the shore in every coastal area, and not at all along the New England coast. You need to research whether cephalopods are commonly found along the costal area you're interested in.

In the UK three species of octopus can be found naturally: *O. vulgaris*, *Bathypolypus arcticus*, and *Eledone cirrhosa*. Chances are that you won't be able to find them around the coasts unless you're out on a fishing boat, where they are sometimes caught on rod and line or in lobster pots. Public aquariums will commonly obtain their display animals from fishermen; the curled or lesser octopus, *E. cirrhosa*, is the most commonly seen species on display. Giant Pacific octopuses are shown from time to time, with *O. vulgaris* being slightly more common.

Along the Southern coasts of the UK you will be able to find cuttlefish eggs from *Sepia officinalis*, as they are occasionally attached to ropes and other objects in the water.

In total there are, at the time of writing this book, sixteen recorded species of cephalopod that can be found around the UK's coasts, but only a few are ever on display at public aquariums.

Diving is a way to get a first-hand look at cephalopods. It's especially attractive for those who already have their SCUBA certification. If not, there are places in the world where you can see cephalopods along the shore or while snorkeling. In the Pacific Northwest part of the U.S., divers can encounter and even interact with giant Pacific octopuses.

Many divers post photos of the cephalopods they've

encountered on diving websites, which is a good place to begin if you're looking for similar experiences. There are even specially arranged tours, such as one for diving with Humboldt squid.

Don't neglect the cephalopods of the past—the fossils in science and natural history museums. If you want first-hand experience with fossils, in many areas you can go on a fossil-hunting trip and with luck actually find some yourself. Also, in a very few special museums, you can find a preserved specimen of a giant squid on display, certainly worth seeing.

If you have an art museum nearby with a good collection of Grecian plates and vases or coins, you could inquire whether the museum has any octopuses or other cephalopods represented on these ancient works of art.

A fossil ammonite, *Placenticeras syrtale*, from the Cretaceous Mancos Shale in eastern Emery County, Utah, about 85 million years old.

There are many possibilities for getting a closer look at cephalopods, in addition to keeping them in your home aquarium. Why not enrich your cephalopod experience by visiting some other species? Or give the fossil exhibit a try. All this will increase your understanding of and appreciation for cephalopods.

Appendix B

Cephalopods in Our History and Culture

Cephalopods have been with us for centuries, and their impact at times seems to exceed that of other inhabitants of our home aquariums. Which other group of marine animals has been more of a source for the inspirations for sea serpents, has served as a source of food even in the time of the ancient Greeks, has featured in various horror movies, and currently is appearing in more advertisements and children's books than you could imagine? Look for cephalopods on stamps, coins, and even in political cartoons—the breadth of their influence is immense. While collecting information for this appendix, we realized that enough material exists to fill an entire book all its own. So this section presents only a brief look at the history of cephalopods and the influences they have had on our

own history. If you're interested in keeping cephalopods, you should know something of this past history to better understand how that octopus in your tank figures into our lives.

EVERYONE LOVES AN OCTOPUS

Well, not exactly everyone, but they appear in various forms in our lives today and have had more influence than any other cephalopod in popular culture. Squids also are influential, often in their largest manifestations, while cuttlefishes and nautiluses both play a lesser role. The argonaut is almost never seen and is largely unknown to the public.

COINS AND POTTERY

If we go back to ancient times, we find many representations of octopuses and sometimes also squids. For instance, there are coins from Mycenae from about 1500 BC that have an octopus on the obverse (face). A realistic octopus appears below. A number of similar coins show other sea creatures, such as crabs, so we are not sure exactly why the octopus is portrayed on the coin. Could the reverse of the coin show the excellent food produced by the region, or is there some more religious significance? We don't know.

The Minoans liked to put octopuses on their pottery to make a free but intricate design. They also liked to depict fish and seaweed, so such pottery is said to be in the Maritime Style. Wonderful flowing octopuses decorate Minoans jugs and plates; many good examples have survived to the present day. The octopuses

This Grecian coin (left) clearly shows an octopus. A small octopus swims alongside the ship on this Roman coin (right, see arrow). Cephalopods are frequent subjects in ancient art but appear only a few times on coins.

Roman Mosaics

The Romans were masters at mosaics, making decorative compositions by arranging small pieces of stone and glass. Their artisans created elaborate mosaics for Roman houses and gardens depicting in a realistic manner a great number of animal species, both terrestrial and marine. In the House of the Faun in the town of Pompeii, preserved by the famed volcanic eruption of Mount Vesuvius, we find mosaics featuring sea life, including an octopus and squid. One beautifully detailed mosaic features an octopus with a large crustacean.

are both realistic and beautiful in the way they occupy the space and their placement on the article to be used.

Jumping forward to Roman times, we find octopuses and other cephalopods and, in fact, many sea creatures represented in the mosaics of the time.

The coins from the same period also show an occasional octopus. On the previous page is one where an octopus accompanies a ship; this octopus is smaller and less realistic than the octopus on the Mycenaean coin. While the Romans no doubt included cephalopods, especially octopuses, on the dinner table, the representation on this coin is of an active octopus in the sea. Note that the octopus is pictured in a way the sailors would never see it, up at the surface with the ship.

SEA SERPENTS AND THE KRAKEN

The kraken is a legendary sea monster often represented as a huge octopus. It was originally reported to live off the coasts of Norway and Iceland, where it generated a dangerous whirlpool as it descended into the ocean's depths. Although the kraken doesn't appear in the early Norse sagas, similar sea monsters are described. Some think that the kraken legend could be based on early sightings of the giant squid. The kraken was supposed to be responsible for sinking ships, putting its huge suckered arms around a vessel and pulling it down to the ocean bottom. It has inspired fear in sailors, and its presence in legend continues to this day.

By the time Linnaeus came out with the first edition of his *Systema Naturae* (1735), in which he classified living organisms, he listed kraken under cephalopods with the name *Microcosmus,* but in later editions he removed the kraken. By the late eighteenth century, we find the kraken depicted as either a large octopus-like creature or more like a giant squid. There was also a belief that a giant octopus existed, so there was more than one huge sea monster to take down ships. Maps of the time often showed sea monsters of various sorts in the oceans. These represented what was believed could be found on the high seas.

CEPHALOPODS IN NATIVE AMERICAN CULTURE

Cephalopods, especially octopuses, have figured in the lives of Native Americans, especially those along the northern Pacific Coast, where they encountered the giant Pacific octopus, *Enteroctopus dofleini.* This octopus was hunted for food and used for fishing bait. Therefore it's not surprising that the octopus found its way into their legends and crafts.

One of the more interesting octopus representations is the octopus bag. This is a bag that ends in eight flaps (or four flaps, each divided into two at the end), resembling an octopus's arms. It was worn around the neck to hold personal possessions. No one knows for sure when this bag was first made or by whom, but good examples of it come from the 1870s and 1880s. They are often heavily embroidered. They can be found from Alaska south along the Pacific Coast of North America and quite far inland, along the fur trading routes. Octopus bags are still made today by Native American artists.

An octopus bag featuring a beautiful floral pattern.

CEPHALOPODS IN THE MILITARY

Ships have been given the names of cephalopods for centuries. Looking at U.S. Navy ships, we find an early submarine named the U.S.S. *Octopus* (1906). There were two submarines called the U.S.S. *Cuttlefish*; one served in World War II. The submarine U.S.S. *Kraken* was launched in 1944 and also saw service in World War II.

The best-known naval vessel with a cephalopod name is the U.S.S. *Nautilus*. A schooner launched in 1799 had that name, and more than 150 years later the world's first operational nuclear submarine bore the name again. This submarine is famous for being the first vessel to pass under the polar ice cap.

The British Royal Navy also has had ships named after a cephalopod, beginning with HMS *Nautilus*, a 16-gun sloop launched in 1762, and the most recent being a submarine launched in 1917.

Cephalopods can also be found on military patches and insignias, where they may represent power or domination.

CEPHALOPOD APPEARANCES IN NINETEENTH- AND TWENTIETH-CENTURY CULTURE

In the latter part of the nineteenth century, cephalopods begin to appear in literature. In 1870 Alfred

Tennyson wrote a poem titled "The Kraken," which most likely influenced the Jules Verne novel *Twenty Thousand Leagues Under the Sea* (1870), in which a giant squid attacks the *Nautilus*, a submarine named after yet another cephalopod. Victor Hugo's novel *Toilers of the Sea* (1866) has a scene in which a worker fights a rather extravagantly described giant octopus—an early use of the octopus as a symbol of power; in this case the Industrial Revolution. And just at the beginning of the twentieth century, Frank Norris published *The Octopus* (1901) a book that describes how railroads (the octopus) try to take over the farmers' land.

Using an octopus to represent grasping power became common in political cartoons in the late nineteenth century and continues into the present time. The menacing and far-reaching octopus has represented just about everything that any group disliked, from various nations (China, Russia, Japan, the U.S.), the Freemasons, city governments, even businesses such as the railroads and Standard Oil.

Classification and Scientific Drawings

The interest in science and in classifying living things that grew in the eighteenth century led to many representations of sea life, including cephalopods. They were dissected and represented in books with wonderfully detailed drawings or engravings. The pages of these books, which are often available for sale today, are used as wall decorations because of their artistic execution.

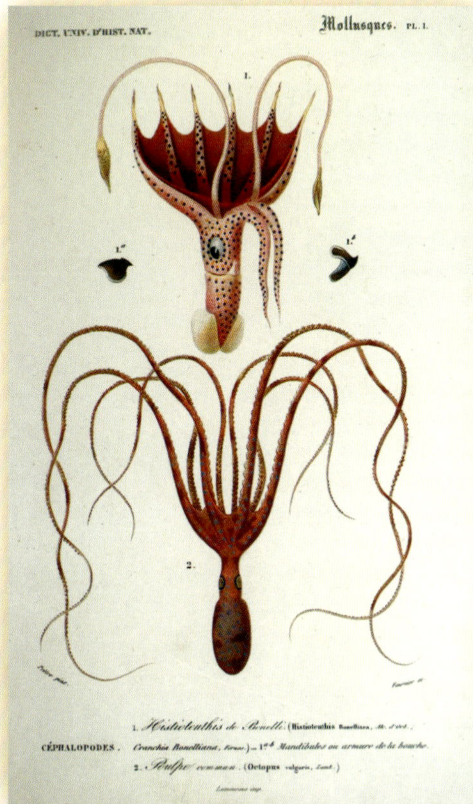

This hand-colored print appeared in a nineteenth-century nature book (*Dictionaire Universel d'Histoire Naturelle* by Charles D'Orbigny, Third Volume, Paris, 1849). These books and their illustrations are collectibles today because of their beauty and skill of execution. *Octopus vulgaris* is recognizable on this page.

Cthulhu

Cthulhu made his first appearance in the 1928 short story "The Call of Cthulhu," by horror writer H.P. Lovecraft. In Lovecraft's hierarchy of beings, Cthulhu was a Great Old One, described as coming from the stars long before man and human history began. Cthulhu and the other Great Old Ones were imprisoned in our world by uncertain methods, and were worshiped and known only by a scattering of cults. However, Cthulhu was described as merely asleep, dreaming of a day when he would return and, in general, wreak havoc on all mankind.

Cthulhu's appearance is said to inspire unspeakable horror and dread, and his body

Great Cthulhu rises from the deep to crush all of man. The cephalopod influence is apparent. This painting is glazed acrylic on canvas, 16 x 20 inches (41 x 51 cm), by artist Greg Ewald.

is variously described as combining that of an octopus and a dragon, as well as having a "pulpy, tentacled head." More specifically, Cthulhu is said to possess the head of a squid or cuttlefish, with tentacles or "feelers" radiating outward from his face.

References to this creature have appeared many times in literature, music, and films up until the present. Cthulhu also appears in cartoon strips and on T-shirts. One of the most recent Cthulhu references is the character Davy Jones in the *Pirates of the Caribbean* film series, who had octopus arms on his head instead of a beard. (He also called upon the kraken to destroy other ships.) The film *Dagon* also belongs to the Cthulhu Mythos—a compilation of themes, settings, characters, and shared elements that Lovecraft and other authors have drawn inspiration from when writing about the Great Old Ones.

Lovecraft's more complete description says: "The cephalopod head was bent forward, so that the ends of the facial feelers brushed the backs of huge fore paws which grasped the croucher's elevated knees." Mention of paws and knees—and elsewhere wings—tells us that Cthulhu is a composite horror, although the most distinctive—and for readers of this book possibly endearing—feature is the squid or cuttlefish head. Certainly filmmakers, writers, and artists have seized on this characteristic with much energy and enthusiasm.

Some musicians make reference to Cthulhu in their songs. Two notable songs from the heavy metal band Metallica which fall into this category are "The Thing That Should Not Be" (off the album *Master of Puppets*) and "The Call of Ktulu" (off *Ride the Lightning*—note the difference in spelling of Cthulhu; "Ktulu" is spelled more phonetically).

Cephalopods in Films and on Television

Cephalopods, or representations of them, have had major roles in many films, where they are generally depicted as an evil monster attacking ships or even setting siege to cities. The best example of this is the 1954 film version of Jules Verne's *Twenty Thousand Leagues Under the Sea* starring Kirk Douglas, James Mason, Paul Lukas, and Peter Lorre.

In general, it is octopuses that play the role of the attacking menace in science fiction and popular films. The octopus filmography is long, so clearly octopuses were, and continue to be, popular villains. One of the most remarkable films was *It Came From Beneath the Sea* (1955), where a giant octopus attacks San Francisco and is seen climbing up tall buildings. Of course, there are the *Octopus* films (*Octopus* and

The Giant Squid

The giant squid, *Architeuthis dux*, appears less in popular culture than the octopus, but it does make an appearance from time to time. Efforts to find the elusive and mysterious giant squid and other huge squids, such as the colossal squid, *Mesonychoteuthis hamiltoni,* have been documented many times on television. The recent filming of an actual live giant squid, first in the larval stage and more recently as an adult, was highly publicized and caused quite a stir in scientific circles as well as with the public and cephalopod enthusiasts at large.

Dead specimens continue to wash up on beaches all over the world and are frequently caught in the haul of fishing boats. They are analyzed and sometimes preserved for viewing in museums and public aquariums. One of the authors of this book, Nancy, has a 3-foot (1-m) long plush giant squid that has served as a mascot for this book, and further inspiration comes from a preserved giant squid beak. Giant squids are now even appearing in cartoons, notably in *Lio*, where a little boy has an imaginary giant squid as a friend.

Size comparison between submature female colossal squid, *Mesonychoteuthis hamiltoni* (bottom), fully mature giant squid, *Architeuthis dux* (top), and semi-mature male teuthologist Steve O'Shea (middle).

Octopus 2: River of Fear, 2000 and 2001, respectively). Both feature a huge octopus, and in the second of these films, the octopus utters growls as it goes about its evil business!

Cephalopods appear in various nature programs and documentaries on television, including documentaries about the search for the giant squid (a popular topic), and several programs on octopuses, cuttlefishes, and squids. Cephalopods, especially octopuses, have also appeared in many episodes of the animated television series *The Simpsons*, sometimes just in the background.

Cephalopods in Science Fiction Literature

By the 1930s pulp-fiction magazines were beginning to appear, including many science fiction titles. Octopuses played a big role, especially in cover illustrations. The octopus or octopus-like creature, huge in size and appearing in any color, battled the heroes. By the 1960s, the covers were more abstract and had more outer-space themes than cephalopod-like creatures, so cephalopod themes became and continue to be rare. The influence of fantasy literature means that today you are more likely to see a dragon-like creature than cephalopod-like being.

Comic books picked up the menacing ceph-like creature, and a threatening octopus repeatedly appears as the bad guy in the story. Here the octopus may be either fairly realistically drawn or depicted more imaginatively. In the *Spiderman* comics, and later the second of the three *Spiderman* films, Spiderman fought a cephalopod-inspired character: Dr. Octopus with his many metallic arms.

Science fiction stories and novels also feature alien creatures with ties to cephalopods. One of the best-

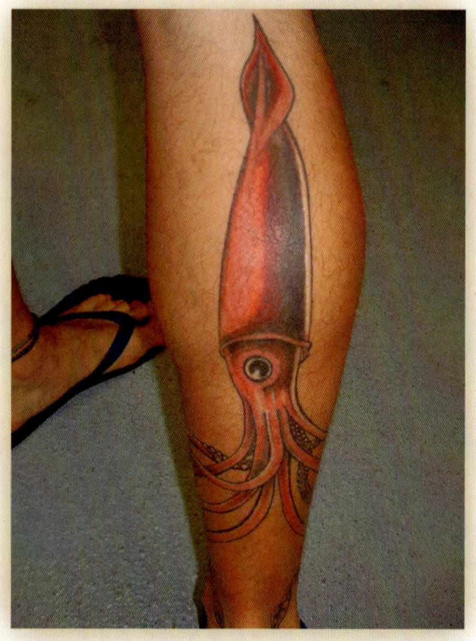

As tattoos have become more popular, it's not surprising that cephalopod tattoos abound. Some octopus designs are chosen because they're attractive or fun or even menacing, but also a number of people who are involved with cephalopods choose an octopus or cuttlefish to embellish an arm, shoulder, back, or leg. Some people prefer a representation of the octopus or cuttlefish they are keeping, while others may choose an early inspiration, like this handsome giant squid tattoo belonging to a marine biologist.

Cephalopods are the inspiration for many plush and plastic toys around the world. (The toys on the right represent fossil cephalopods.)

known novels is *The Kraken Wakes* by John Wyndham (1953), in which an alien intelligence invades Earth, hiding deep within the sea. The title refers to the destruction wrought by the invader, not a cephalopod-like form. And of course the many Lovecraft stories and continuation of the Cthulhu Mythos by other authors would be included in cephalopod science fiction and horror stories.

CEPHALOPODS ARE MORE VISIBLE IN OUR LIVES

Any mention of cephalopods in our culture would have to include the many appearances in the advertising of various products. They are presented in a positive light, helping to sell products or help decorate your home. Although you may not be specifically watching for them, octopuses appear, often as cartoons, to sell products such as air fresheners. The octopus shape can make a product more attractive, such as the octopus-shaped back massager.

We just mentioned the plush giant squid, but did you know you can even buy a plush cuttlefish or plush Cthulhu? Nautilus shells and artwork featuring them are often used in home decorations and appear as an almost abstract design on dishes and on wall art. Octopuses and nautiluses are the cephalopod shapes most seen in jewelry. Children seem especially taken with octopuses; the number of toys and books featuring octopuses, real or imaginary, is amazing. (Many octopus toys are, however, anatomically incorrect, usually having two funnels rather than the single funnel a real octopus has.)

RESOURCES

Magazine

Tropical Fish Hobbyist
1 TFH Plaza
3rd & Union Avenues
Neptune City, NJ 07753
E-mail: info@tfh.com
www.tfhmagazine.com

Internet Resources

Aquaria Central
www.aquariacentral.com

Aquarium Hobbyist
www.aquariumhobbyist.com

The Cephalopod Page
www.cephalopodpage.org

CephBase
www.cephbase.org

Marine Aquarium Advice
www.marineaquariumadvice.com

Microcosm Aquarium Explorer
www.microcosmaquariumexplorer.com

National Resource Center for Cephalopods (NRCC)
www.cephalopod.org

Reef Central
www.reefcentral.com

TFH **Magazine Forum (formerly Tropical Resources)**
http://forums.tfhmagazine.com

TONMO (The Octopus News Magazine Online)
www.tonmo.com

Wet Web Media
www.wetwebmedia.com

A World of Fish
www.aworldoffish.com

Associations and Societies

Federation of American Aquarium Societies (FAAS)
E-mail: Jbenes01@yahoo.com
www.faas.info

Marine Aquarium Council (MAC)
E-mail: infor@aquariumcouncil.org
www.aquariumcouncil.org

Resources

Books

Calfo, Anthony and Robert Fenner. 2003. *Invertebrates: An Essential Guide to Selection, Care and Compatibility.* Natural Marine Aquarium Series, Vol. 1. Reading Trees, United States.

Cousteau, Jacques-Yves and Philippe Diol . 1973. *Octopus and Squid.* Doubleday, Garden City, New York, United States.

Ellis, Richard. 1994. *Monsters of the Sea.* Lyons Press, Connecticut, United States.

Fenner, Robert M. 2008. *The Conscientious Marine Aquarist, Second Edition.* Microcosm Ltd./TFH Publications, Inc., Neptune City, New Jersey, United States.

Foss, Svein A. and A.J. Nilsen. 1996. *The Modern Reef Aquarium*, Birgit Schmettkamp Verlag, Bornheim, Germany.

Hunt, James C. 1996. *Octopus and Squid.* Monterey Bay Aquarium, Monterey, California, United States.

Lovecraft, H.P. 2005. *Tales.* Library of America, New York, United States.

Monks, Neale. and Philip Palmer. 2002. *Ammonites.* Smithsonian Institution Press, Washington D.C., United States.

Moynihan, Martin. 1985. *Communication and Noncommunication by Cephalopods*, Indiana University Press, Indiana, United States

Norman, Mark. 2000. *Cephalopods: A World Guide.* Conch Books, Hackenheim, Germany.

Peletta, Michael S. 2001. *The New Marine Aquarium.* Microcosm Ltd./TFH Publications, Inc., Neptune City, New Jersey, United States.

Sprung, J. and C. Delbeek. 1997. *The Reef Aquarium*, Vol. 1-3, Ricordea Publishing, Coconut Grove, Florida, United States.

Ward, Ashley. 2008. *Questions & Answers on Saltwater Aquarium Fishes.* TFH Publications, Inc., Neptune City, New Jersey, United States.

BIBLIOGRAPHY

Books

Boyle, P. 1991. *UFAW Handbook on the Care and Management of Cephalopods in the Laboratory*, Universities Federation for Animal Welfare (UFAW). Hyperion Books, UK

Boyle, Peter and Paul Rodhouse. 2005. *Cephalopods: Ecology and Fisheries*. Blackwell, Oxford, UK.

Clarkson, Enk. 1998. 4th edition. *Invertebrate Paleontology and Evolution*. Blackwell, Oxford, UK.

Erhardt, H. and H.Moosleitner. 1998. *Baensche Marine Atlas*, Vol. 3, Invertebrates Microcosm/ Mergus, Vermont, United States.

Hanlon, Roger T. and John B. Messenger. 1996. *Cephalopod Behaviour*. Cambridge University Press, Cambridge, UK.

Monks, Neale. and Philip Palmer 2002. *Ammonites*. Smithsonian Institution Press, Washington D.C., United States.

Moynihan, Martin. 1985. *Communication and Noncommunication by Cephalopods*. Indiana University Press, Bloomington, Indiana, United States.

Nesis, Kir. *Cephalopods of the World: Squids, Cuttlefishes, Octopuses, and Allies*. 1987. Translated from the Russian by B.S. Levitov. TFH Publications, Inc., Neptune City, New Jersey, United States.

Norman, Mark. 2000. *Cephalopods: A World Guide*. Conch Books, Hackenheim, Germany.

Wells, M.J. 1978. *Octopus: Physiology and Behaviour of an Advanced Invertebrate*. Chapman and Hall, London, UK.

Bibliography

Articles

Anderson, R. 1992. 20,000 Tentacles under the Sea: Cephalopods in cinema. Of Sea and Shore, 15 (2): 78-84 Updated version (1999) on www.thecephalopodpage.org.

Anderson, R.C. and J.B. Wood. 2001. Enrichmentfor giant Pacific octopuses: happpy as a clam? *Journal for Applied Animal Welfare Science* 4(2): 157-168.

Anderson, Roland C. 2008 Novel Foods as Enrichment for Giant Pacific Octopuses (GPOs) Drum and Croaker 39 (2008) 62-64.

Caldwell, Roy. Death in a Pretty Package: The Blue-Ringed Octopus. *Freshwater and Marine Aquarium Magazine*, 23 (3):8-19. Also available on The Cephalopod Page, wwwthecepahalopodpage.com.

Caldwell R. and Shaw C.D. 2002. Mimic Octopuses: Will we love them to death? The Cephalopod Page www.thecephalopodpage.org.

Crook, R. and Basil, J. 2008. A biphasic memory curve in the chambered nautilus, Nautilus pompilius L. (Cephalopoda: Nautiloidea). *J. Exp. Biol.* 211, 1992-1998.

Daw, A. and Gregory Jeff Barord. "Aquarium Science: Husbandry of the Nautilus: Aspects of its Biology, Behavior, and Care", *Tropical Fish Hobbyist*, Vol. LV, No. 10 (#615), June 2008.

Huffard, CL. 2007. Ethogram of *Abdopus aculeatus* (Cephalopoda: Octopodidae): can behavioral characters inform octopodid taxonomy and systematics: *J Molluscan Studies* 73:185-193.

Huffard, CL. 2006. Locomotion by *Abdopus aculeatus* (Cephalopoda: Octopodidae): walking the line between primary and secondary defenses. *Journal of Experimental Biology* 209: 3697-3707.

Huffard, Christine L, Roy Caldwell, and Farnis Boneka. 2008. Mating behavior of *Abdopus aculeatus* (d'Orbignuy, 1834) (Cephalopoda: Octopodidae) in the wild, *Marine Biology*, Vol. 154, No. 2: 353-363.

Mather, Jennifer M and Roland C. Anderson. Ethics and invertebrates: a cephalopod perspective, *Diseases of Aquatic Organisms*, Vol . 75: 119-129, 2007.

Monks N. 2003. A Broad Brush History of the Cephalopoda. The Cephalopod Page wwwthecephalopodpage.org.

Robinson, A. 2000. "Octopus briareus" (Online), Animal Diversity Web. Accessed June 05, 2008 at http://animaldiversity.ummz.umich.edu/site/accounts/information/Octopus_briareus.html.

Ross, Richard. The Ballet of the Wunderpus. C *the Journal of Aquatic Science, Travel and Adventure.* 2007 Vol. 1, No. 4: 65-71.

Wood J.B. 1995. Raising and Rearing Octopus briareus. First published in: The Breeder's Registry; *The Journal of Maquaculture.* vol 3, no 2. The Cephalopod Page www.thecephalopodpage.org.

Wood J.B. and Wood D.A. 1999. Enrichment for an Advanced Invertebrate (in Adobe Acrobat format) This article originally appeared in The Shape of Enrichment 8(3): 1-5.

Note: **Boldfaced** numbers indicate illustrations; an italic *t* indicates tables.

A

Index

Index

Photo Credits

Gregory J. Barord: 164, 220

Big Zen Dragon (Shutterstock): Title Page

Kat Bolstad: 219

Kevin Bylund: 211 (top)

Roy Caldwell: 6, 13, 31, 43, 52, 59, 96, 123, 124, 128, 130, 132-133, 142, 144, 146-147, 153, 155, 181, 185, 190, 205, 208, 210

Dr. Bruce Carlson: 65

cbpix (Shutterstock): 211 (bottom)

Dominick Cirigliano: 10, 54

Robyn Crook: 170, 171, 172

Adam Daw: 168, 169

Colin Dunlop: 27, 61, 82, 83, 84, 100, 110, 115, 116, 118, 120, 154, 176, 178, 183, 184, 189, 195, 204, 221 (left)

EcoPrint (Shutterstock): 25

Arthur W. Erickson: 216

Greg Ewald: 218

Phil Eyden: 18, 20, 221 (right)

James Fatherree: 8, 33, 94

Curt Gerken: 38

Michael Gilroy: 34

Ilya D. Gridnev (Shutterstock): Cover

Markus Holcomb: 72

Christine Huffard: 4, 22, 138, 141, 158, 173

Ray Hunziker: 111

Dan Hyslop: 76 (left & right), 93

Stelian Ion (Shutterstock): 30

David Kelly (Shutterstock): 212

Bill Kennedy (Shutterstock): 80

Bill King: 119, 214 (left & right), 217

Nancy King: 14 (top & bottom), 107, 148, 196, 198, 199

Jeff Kurtz: 58

Herve Lavigne (Shutterstock): 187

James Lee: 32

Oliver Lucanus: 113

Johnny Lye (Shutterstock): 3

Matthew McGhee: 44

Dermot McMahon: 57, 69, 70, 71, 73, 74

MP. & C. Piednoir: 165, 238

Daniel Pon: 5, 28, 41, 42, 50, 66, 121 (left & right), 135, 201

Tom Robbrecht (Shutterstock): 40

Richard Ross: 108, 136, 157, 160, 162, 166, 193, 194, 203

Michael Roth: 90

Dario Sabljak (Shutterstock): 137

Jennifer Saitz: 163

Carol Sauer-Corwin: 103, 143, 149

Rena Schild (Shutterstock): 95

Michael Seidman: 47, 49, 224

Zyan Silver: 88, 102, 131, 134

Dr. Dwight Smith: 174, 237

Mark Smith: 75

Joe Strickland: 19, 24, 92, 98, 104, 126, 179, 223

TFH Photo Archives: 207

Dale Tyler: 87, 151

velefante (Shutterstock): 79

D Whatley: 63, 77, 122

Barry Wisebram: 156, 161